Machine Learning
Theory to Applications

Seyedeh Leili Mirtaheri

Assistant Professor, Electrical and Computer Engineering Department
Kharazmi University, Tehran

Reza Shahbazian

Electrical Engineering Research Group
Faculty of Technology and Engineering Research Center
Standard Research Institute, Alborz and
Department of Mathematics and Computer Science
University of Calabria, Italy

T0199645

CRC Press
Taylor & Francis Group
Boca Raton   London   New York

CRC Press is an imprint of the
Taylor & Francis Group, an **informa** business

A SCIENCE PUBLISHERS BOOK

First edition published 2022
by CRC Press
6000 Broken Sound Parkway NW, Suite 300, Boca Raton, FL 33487-2742

and by CRC Press
4 Park Square, Milton Park, Abingdon, Oxon, OX14 4RN

© 2022 Taylor & Francis Group, LLC

CRC Press is an imprint of Taylor & Francis Group, LLC

Library of Congress Cataloging-in-Publication Data (applied for)

ISBN: 978-0-367-63453-7 (hbk)
ISBN: 978-0-367-63456-8 (pbk)
ISBN: 978-1-003-11925-8 (ebk)

DOI: 10.1201/9781003119258

Typeset in Times New Roman
by Radiant Productions

Preface

The motivation to write this book is based on famous poetry from a great Persian poet that says: "However much science thou mayest acquire Thou art ignorant when there is no practice in thee."

This poem emphasizes the practical aspects of knowledge as we try to present in this book. Science and knowledge are windows to the world of progress and consciousness, and if they become operational, they can put the man on the wing of the miracle of comfort and reach. All the tools and instruments we use now cause a fundamental difference between the quality of life in the past and the feature resulting from the efforts.

Today, machine learning has found wide applications in various industries, such as recommender systems and customer behavior analyzing, diagnosing diseases, detecting fraud and violations in banking and insurance, stock price forecasting, detecting suspicious objects in CCTV films, and many other fields. Extensive applications of machine learning in various fields have led us to look at this issue from a practical point of view and to write a book titled Machine Learning: Theory to Applications.

In this book, we start by discussing the general concepts of artificial intelligence (AI), machine learning (ML), deep learning (DL), training, reinforcement learning, autoencoders, supervised, semi-supervised and unsupervised learning. Then we introduce a set of basic operations in the matrix form to clarify machine learning concepts in detail. We continue with the idea of machine learning (ML) and the study of computer algorithms that improve automatically through data usage. For the practical approach, we addressed the issue of how to split the dataset in train, validation, and test and how to check the performance of a trained model. We discuss the standard evaluation metrics, including the F1-Score, accuracy, precision, and recall. We introduce over and underfitting concepts and Lasso, Ridge, and Dropout regression as regularization techniques. The ceiling analysis is presented as a great technique to debug machine learning-based systems. Because of the critical state of deep learning in solving today's problems as a branch of machine learning algorithms based on artificial neural networks, we discuss the deep learning networks and models in detail along with the multi-layer perceptrons (MLPs), Radial Basis Function Networks (RBFNs), Restricted Boltzmann Machines (RBMs) and Autoencoders. The variational autoencoder (VAE) and Generative adversarial networks (GANs) can produce artificial data samples based on a limited collected data set. We discuss VAE and GAN as a separate section in this book. Finally, we present a recent time-slide inclusive overview with collations and trends in developing and using brand-new Artificial Intelligence software. We also provide an overview of huge parallelism support capable of scaling computation effectively and efficiently in the epoch of Big Data.

We hope that this book is widely read and the readers enjoy it while step into the new world of experience.

Contents

Abbreviations

AI	Artificial intelligence
ML	machine learning
RL	reinforcement learning
MLP	multi-layer perceptron
MDP	Markov decision process
CAA	crossbar adaptive array
ANN	Artificial Neural Networks
SVMs	Support vector machines
DAG	directed acyclic graph
GA	Genetic Algorithm
FL	Federated learning
DL	Deep learning
CNNs	convolutional neural networks
UAT	universal approximation
PT	probabilistic theorem
CDF	cumulative distribution function
MSE	mean squared error
DNN	deep neural network
GANs	generative adversarial networks
RNNs	Recurrent neural networks
LSTMs	Long short term memory networks
RBFNs	Radial Basis Function Networks
MLP	Multi-layer perceptrons
DBNs	Deep belief networks
RBMs	Restricted Boltzmann Machines
VAE	variational autoencoder
GPUs	Graphics Processing Units

FPGA	Field Programmable Gate Array
TPU	Tensor Processing Unit 3.0
API	Application Programming Interface
LR	linear regression
VW	Vowpal Wabbit
SGD	stochastic gradient descent

CHAPTER 1
Introduction

Artificial intelligence (AI) attempts to build intelligent entities and understand how they perceive, understand, and predict. The dream of creating machines that can understand and think goes back to ancient Greece. Years before the existence of programmable computers, it was hard for people to imagine such a thing. AI is a prospered field that has many applications in today's society and active research topics. Researchers look at the potential uses of intelligent software in medical diagnosis, understanding speech or images, and automating routine labor.

At the beginning of AI's emergence, the field rapidly solved problems described by mathematical rules that are intellectually difficult for humans but relatively simple for computers. AI solves the tasks that are easy to perform by humankind, the kind of problems that we solve intuitively, that feel automatic, like recognizing spoken words or faces in images.

Many of the early AI successes did not require computers to have much knowledge about the world, such as overcoming world champion Garry Kasparov with IBM's Deep Blue chess-playing system in 1997. It is interesting to know that some conceptual tasks, which are some of the most difficult ones for a human being, are the computer's most straightforward tasks. A person's everyday life requires a lot of subjective and intuitive knowledge about the world that is difficult to articulate formally. A computers' intelligent behavior needs to capture this same informal knowledge, but this is critical in artificial intelligence. Several AI projects seek to encode knowledge of the world into formal languages. A computer can reason about statements automatically using logical inference rules, also known as the knowledge base manner.

The ability to access systems knowledge by extracting patterns from raw data, known as machine learning (ML), is a problem that systems face based on hard-coded knowledge. Machine learning allowed computers to solve problems involving knowledge of the real world and make decisions that appear subjective. The performance of machine learning algorithms depends heavily on the representation of the given data. For example, to recommend cesarean delivery by logistic regression, the AI system does not examine the patient directly. The doctor provides the system with several pieces of relevant information known as features, such as the presence or absence of a uterine scar. Logistic regression learns how each of these features of the patient is related to different results. However, it cannot influence the way that these features are defined. Because of the small association of individual pixels in MRI scans with any complications during labor, if logistic regression is given to the patient by MRI scan, rather than a formal physician report, an erroneous prediction may be made. Data representation dependency is a general phenomenon that appears throughout computer science and even daily life.

Machine learning refers to a vast set of tools, classified as supervised, unsupervised, and semi-supervised for understanding data. Generally, supervised machine learning involves building a statistical model for predicting or estimating an output based on one or more inputs. Problems in various fields, including medicine, astrophysics, business, and public policy, fit into this machine learning class. In unsupervised machine learning, we can learn relationships and structure from such data, and there is no supervising outcome from existing inputs. Some examples of such statistical learning applications are as follows [1]:

- Predicting the possibility of a heart attack recurrence for someone who has been hospitalized once for a heart attack.
- Predicting the price of a stock in 3 months from now, based on economic data.
- Identifying the numbers in a hand-written ZIP code from a digitized image.
- Estimating the blood glucose of a diabetic person from the infrared absorption spectrum of that person's blood.
- Identifying the risk factors for cancer based on clinical and demographic variables.

This book is about learning from data that plays a crucial role in statistics, data mining, and artificial intelligence, intersecting with engineering disciplines. In a typical scenario, we have an outcome measurement to predict based on a set of features (such as diet and clinical measures). We have a training (learning) set of data, in which we observe the outcome and feature measurements for a collection of objects (such as people). Using this data and intending to predict new things, we create a prediction model or learner. A good learner can expect such an outcome accurately.

The examples above describe the supervised learning problem. In the unsupervised learning problem, only the features are observed and these features have no measurements of the outcome. It is our job as machine learning experts to describe the organization or clustering of the data.

Consider the example of recognizing hand-written digits, illustrated in Figure 1.1. Each integer corresponds to a 28 × 28-pixel image, and a vector $\mathbf{x}$ comprising 784 real numbers represents the digit. The goal is to build a machine that will take such a vector $\mathbf{x}$ as input and return one of the digits 0, 1,..., 9 as output. Given the wide variability of handwriting, this is an intractable problem. Handcrafted rules or heuristics can be used for distinguishing the digits based on the shapes of the strokes, but in practice, this manner leads to an increase in regulations and will have poor results.

Using a machine learning approach in which the training set is a large set of N digits $\{\mathbf{X}_1, ...,\mathbf{X}_N\}$ to tune the parameters of an adaptive model, we can obtain better results. The division of digits in the training set is already known, typically by inspecting them individually and labeling them by hand. We can express the category of a digit using a target vector $\mathbf{t}$ for each digit image $\mathbf{x}$, representing the corresponding digit's identity. Suitable techniques for classification will be discussed later in this book.

The function y(x) is considered the output obtained from the machine learning algorithm in which y is a generated output vector corresponding to the new x-digit input and is encoded in the same way as the target vectors. The exact shape of the function $\mathbf{y}(\mathbf{x})$ is determined based on the training phase's data. After training the model, it can identify new digital images that contain a set of tests.

Figure 1.1. Some instances of hand-written digits.

Generalization is the ability to categorize the new examples absent in the training phase correctly. In practical applications, the input vectors' variability will be such that the training data can only form a small part of all possible input vectors. Therefore, generalization is the main goal in pattern recognition. In most practical applications, the original input variables are typically pre-processed and transformed into some new space of variables to solve the problem easier. For example, in the digit recognition problem, the translation and scaling of digit images have a fixed size for each digit and thus reduce the variability within each class. This pre-processing stage is sometimes also called feature extraction. Note that new test data must be pre-processed using the same steps as the training data [2].

Pre-processing might also be performed to speed up computation. For example, in real-time face detection with a high-resolution video stream, the computing device handles vast numbers of pixels per second. Presenting these directly to a complex pattern recognition algorithm may be computationally infeasible. Therefore, there is a need to find useful features as the input of the pattern recognition algorithm that contains useful discriminatory information to distinguish faces from non-faces and be fast computable. For example, for rapid face recognition, the average value of the image intensity in a rectangular subregion can be evaluated [3]. So, we have a dimensionality reduction because of being smaller the number of such features than the number of pixels. During pre-processing, the information often discarded is not essential, and the overall accuracy of the system should not be compromised.

As mentioned earlier, applications in which the training data are composed of the input vectors and their corresponding target vectors form supervised learning problems. Classification supervised learning is to identify a new observation that belongs to which of a set of categories, for example, assigning each input vector to one of a finite number of discrete categories in the digit recognition example. If the desired output consists of continuous variables (for instance, prediction of the yield in a chemical manufacturing process), then the task is called regression.

The training data consists of a set of input vectors **x** without any or limited corresponding target values in other pattern recognition problems. Trying to discover similar examples within the data in such situations is called clustering and is a goal in unsupervised learning. Determining the distribution of data in the input space, known as density estimation, or projecting data from a high-dimensional space into two or three dimensions for visualization are other objectives of an unsupervised learning problem.

The technique of reinforcement learning (RL) [4, 5] is concerned with the problem of finding suitable actions to take in a given environment to maximize a reward and is one of the basic machine learning models, along with supervised learning and unsupervised learning. Reward functions describe how the agent should behave and allow AI platforms to arrive at conclusions instead of arriving at a prediction. RL algorithm does not need labeled input/output pairs, in contrast to supervised learning, but needs to discover examples of optimal outputs through trial and error. For instance, using RL techniques, a neural network can learn to play backgammon to a high standard [6] in which the network must learn to take a board position as input by using the result of a dice throw and produce a perfect move as the output. This is done by playing many games against itself to reinforce what is learned. A significant challenge is that a backgammon game can involve dozens of moves; however, the reward is only achieved at the end of the game, in the form of a victory. Then, the reward must be attributed appropriately to all the movements (including good ones and others less so) associated with it. This is named a credit assignment problem. RL's general feature is the trade-off between exploration (aiming to examine the impact of new types of actions) and exploitation (due to high access reward using actions known). Too much focus on each of them will lead to poor results.

We can solve many artificial intelligence tasks by designing the right set of features to extract and then by providing these features to machine learning algorithms. However, it is difficult to know what features should be extracted. For example, assume that we want to write a program to identify cars available in the images. We know that cars have wheels, and the presence of a wheel can be used as a feature. However, it is difficult to accurately describe the wheel in terms of pixel values because of the possibility of image complexity due to shadows falling on the wheel.

One solution to this problem is to use machine learning to discover both the mapping from representation to output and the representation itself. This approach is known as representation learning (or feature learning). Learned representation often results in a much better performance than can be obtained with manual-designed representations and allows AI systems to adapt to new tasks rapidly. Unlike the manual design of features which requires a lot of time and effort for humans, the representational learning algorithm can detect a good set of features almost in no time.

Autoencoder is a clear example of a representation learning algorithm; it combines an encoder and a decoder function. The encoder function converts the input data into a different representation, and the decoder converts the new representation back into the original format. Different kinds of autoencoders aim to achieve different types of properties. In designing features or algorithms for learning features, we usually separate the various factors that explain the observed data. In this context, we use the word "factors" simply to refer to separate sources of influence; the factors are usually not combined by multiplication and are often not directly observed quantities. Instead, they may exist either as unobserved objects or unobserved forces in the physical world.

A challenge in many real-world AI applications is that many of the factors of variation impress every single piece of data we can observe. The single pixels in two given images of a red car and a black one might be very close at night. A side view of a car depends on the viewing angle. In most applications, we need to separate various features from each other and discard negligible ones. It is not easy to extract such features from naive data. It is hard to do so to have high-level abstract features. Many factors of variation, such as a speaker's accent, can be detected only using an experienced, nearly human-level understanding of the data. It is as tricky as obtaining a representation to solve the original problem. Therefore, at first, representation learning does not seem to help us.

Deep learning has a solution for this crucial problem in representation learning by introducing more straightforward representations. Deep learning allows the computer to develop complex concepts out of simpler ones. A deep learning system can represent complex concepts from more straightforward concepts such as corners and contours in an image of a person. These features are defined in terms of edges sequentially. The principal example of a deep learning model is the feed forward deep network or multi-layer perceptron (MLP). A multi-layer perceptron is just a mathematical function mapping some set of input values to output values. Composing many simple functions builds this function. We can assume each application of a different mathematical function as a new representation of the input.

Learning the proper representation for the data provides a view on deep learning. Secondly, the computer can learn a multi-step algorithm with the depth that we have in deep learning. We can think of each layer of the representation as to the computer's memory in which another set of instructions is executed in parallel. More instructions can execute sequentially in more deep networks. There is great power in sequential executions because later instructions can refer back to earlier instructions. According to this view, all the information in a layer's activations is not necessarily encoded in variations that explain the input. The representation also keeps the information of the states that helps to execute a program to make sense of the input. This information could be equivalent to a counter or pointer in a program. The input content cannot be used to do something specific, but it helps organize the model's processing.

There are two ways to measure the depth of a model. The first one is based on the number of sequential instructions that are needed for architecture evaluation. We can consider it the same as the length of the longest path through a flow chart of the model for computing the outputs of a model given its inputs. In the flowcharts of two equivalent programs, we may have different lengths, and it depends on the languages the programs are written in. Similarly, we can have additional depth in two flowcharts drawn for the same function. It depends on which parts we allow to be used as individual steps in the flowchart.

In the second approach used by deep probabilistic models, the depth of the model is not the flowchart of the computing program but the depth of the graph describing the relation between concepts. The system's understanding of a single concept may refine given information of more complex concepts. Thus, the depth of the program's flowchart for computing the representation of each concept may be much deeper than the graph of the concepts themselves. For example, if there is an image of a face with one eye in shadow, at first, an AI system may see only one eye. Then, by detecting a face's presence, it can figure that a second eye is probably present as well. In this case, there is just a layer for the eyes and one for the face in the graph of concepts. There are only two layers, but if we refine our estimate of each concept given the other n times, the graph of computations includes 2n layers.

It is not always clear which view is the most related one—the depth of the computational graph or the depth of the probabilistic modeling graph, and there is a different choice for a set of most minor elements to constructing the graph. Therefore, there is no single correct value for the depth of architecture and no single valid value for the length of a computer program. There is neither an agreement about how much depth a model needs to certify as "deep." However, deep learning can cautiously be interpreted as models with a large structure of learned functions and learned concepts that traditional machine learning does.

In short, deep learning is an approach to AI. Machine Learning is a computer system technique to develop an enhancement by experience and data, and AI is a type. For the real-world environment that is so complicated, machine learning is the only feasible approach to develop AI systems that can operate in this environment. In deep learning, as a specific kind of machine learning, we achieve great power and flexibility by learning to represent the world as a nested hierarchy of concepts. Figure 1.2 illustrates the relationship between these different AI disciplines.

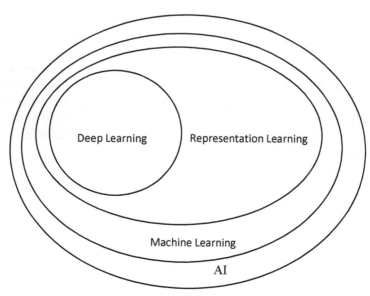

Figure 1.2. The relationship between these different AI disciplines.

CHAPTER 2
Linear Algebra

IIIIIIIIIIIIIIIIIIIIIIIIIIIIIIIIIIIIIIIIIIIIIIIIIIIIIIIIIIIIIIIIIIIIIIIIIIIIIIIIIIIIIIIIIIIIIIIIIIIIIIIIIIIIIIIIIIIIIIIIIIIIIIIIIIIIIIIIIIIIIIIIIIIIIIIIIIIIIIIIIIIIIIIII

This chapter aims to introduce a set of basic operations in matrix form to clarify machine learning concepts in detail. In mathematics, linear algebra is a branch that focuses on linear equations such as:

$$a_1 x_1 + \cdots + a_n x_n = b$$

where a_i is the i'th coefficient for the i'th variable x_i, and b is a constant value. Linear algebra, including linear operations and matrix calculations, are the foundation of many non-linear machine learning tasks. The equation below gives a matrix $\mathbf{C}$ with N rows and M columns.

$$\mathbf{C} = \begin{bmatrix} c_{11} & c_{12} & \cdots & c_{1M} \\ c_{21} & c_{22} & \cdots & c_{2M} \\ \vdots & \vdots & \ddots & \vdots \\ c_{N1} & c_{N2} & \cdots & c_{NM} \end{bmatrix}$$

First, we start with the definition of matrix addition which is as follows:

$$\mathbf{C} + \mathbf{D} = \begin{bmatrix} c_{11} + d_{11} & c_{12} + d_{12} & \cdots & c_{1M} + d_{1M} \\ c_{21} + d_{21} & c_{22} + d_{22} & \cdots & c_{2M} + d_{2M} \\ \vdots & \vdots & \ddots & \vdots \\ c_{N1} + d_{N1} & c_{N2} + d_{N2} & \cdots & c_{NM} + d_{NM} \end{bmatrix}$$

In matrix operations, there are different kinds of products. Here, we explain some sorts of products that are primarily used in machine learning applications. The regular multiplication of two matrices $\mathbf{C} \in \mathbb{R}^{N \times M}$ and $\mathbf{D} \in \mathbb{R}^{M \times K}$ is defined as $\mathbf{CD} \in \mathbb{R}^{N \times K}$ if the columns of the first matrix and the rows of the second matrix are the same. The equation below shows the regular multiplication of two matrices

$$\mathbf{CD} = \begin{bmatrix} c_{11}d_{11} + c_{12}d_{21} + \cdots + c_{1M}d_{M1} & c_{11}d_{12} + c_{12}d_{22} + \cdots + c_{1M}d_{M2} & \cdots & c_{11}d_{1K} + c_{12}d_{2K} + \cdots + c_{1M}d_{MK} \\ c_{21}d_{11} + c_{22}d_{21} + \cdots + c_{2M}d_{M1} & c_{21}d_{12} + c_{22}d_{22} + \cdots + c_{2M}d_{M2} & \cdots & c_{11}d_{1K} + c_{22}d_{2K} + \cdots + c_{2M}d_{MK} \\ \vdots & \vdots & \ddots & \vdots \\ c_{N1}d_{11} + c_{N2}d_{21} + \cdots + c_{NM}d_{M1} & c_{N1}d_{12} + c_{N2}d_{22} + \cdots + c_{NM}d_{M2} & \cdots & c_{N1}d_{1K} + c_{N2}d_{2K} + \cdots + c_{NM}d_{MK} \end{bmatrix}$$

On the other hand, the element wise multiplication is defined if the dimensions of two matrices are the same and can be written as follows:

$$\mathbf{C} \circ \mathbf{D} = \begin{bmatrix} c_{11}d_{11} & c_{12}d_{12} & \cdots & c_{1M}d_{1M} \\ c_{21}d_{21} & c_{22}d_{22} & \cdots & c_{2M}d_{2M} \\ \vdots & \vdots & \ddots & \vdots \\ c_{N1}d_{N1} & c_{N2}d_{N2} & \cdots & c_{NM}d_{NM} \end{bmatrix}$$

The Kronecker product is another matrix multiplication for two matrices $\mathbf{C} \in \mathbb{R}^{N \times M}$ and $\mathbf{D} \in \mathbb{R}^{S \times K}$ which is defined as $\mathbf{C} \otimes \mathbf{D} \in \mathbb{R}^{NS \times MK}$ and written as follows:

$$\mathbf{C} \otimes \mathbf{D} = \begin{bmatrix} c_{11}\mathbf{D} & c_{12}\mathbf{D} & \cdots & c_{1M}\mathbf{D} \\ c_{21}\mathbf{D} & c_{22}\mathbf{D} & \cdots & c_{2M}\mathbf{D} \\ \vdots & \vdots & \ddots & \vdots \\ c_{N1}\mathbf{D} & c_{N2}\mathbf{D} & \cdots & c_{NM}\mathbf{D} \end{bmatrix}$$

The transpose operation is another valuable and essential matrix operation that changes the rows and columns of a matrix. Consider a matrix $\mathbf{C}$ with N rows and M columns, then the transpose of the $\mathbf{C}$ matrix is indicated by $\mathbf{C}^T$, which is defined as follows:

$$\mathbf{C}^T = \begin{bmatrix} c_{11} & c_{21} & \cdots & c_{N1} \\ c_{12} & c_{22} & \cdots & c_{N2} \\ \vdots & \vdots & \ddots & \vdots \\ c_{1M} & c_{2M} & \cdots & c_{NM} \end{bmatrix}$$

The Hermitian operation is considered as the complex version of transpose and can be written as follows:

$$\mathbf{C}^H = (\mathbf{C}^*)^T$$

where * is the conjugate operator.

Matrix rules

The inverse of matrix C is defined if it has equal rows and columns and a non-zero determinant $\det(\mathbf{C}) \neq 0$. Note that C and its inverse, which is represented by, $\mathbf{C}^{-1}$ is equivalent to identity matrix I $\mathbf{C}\mathbf{C}^{-1} = \mathbf{I}$. As its name implies, the identity matrix has diagonal elements equal to 1 and non-diagonal elements equal to 0. Now that the reader has a better intuition on basic mathematical operations in the matrix form, we get into explaining matrix rules as follows:

$$(\mathbf{C}+\mathbf{D})^T = \mathbf{C}^T + \mathbf{D}^T$$

$$(\mathbf{C}+\mathbf{D})^H = \mathbf{C}^H + \mathbf{D}^H$$

$$(\mathbf{CD})^{-1} = \mathbf{D}^{-1}\mathbf{C}^{-1}$$

$$(\mathbf{CD})^T = \mathbf{D}^T\mathbf{C}^T$$

$$(\mathbf{CD})^H = \mathbf{D}^H\mathbf{C}^H$$

$$(\mathbf{C}^T)^{-1} = (\mathbf{C}^{-1})^T$$

$$(\mathbf{C}^H)^{-1} = (\mathbf{C}^{-1})^H$$

In the above equations, T and H represent transpose and Hermitian operations, respectively.

The trace of a matrix is defined as a square matrix and represented by Tr(C). Tr(C) is equal to the summation of diagonal elements of the matrix C. Below, some of the essential rules for trace operations are presented:

$$\text{Tr}(\mathbf{C}) = \sum_i c_{ii}$$

$$\text{Tr}(\mathbf{C}) = \sum_i \lambda_i, \text{ where } \lambda_i = i^{th} \text{ eigenvalue}(\mathbf{C})$$

$$\text{Tr}(\mathbf{C}) = \text{Tr}(\mathbf{C}^T)$$

$$\text{Tr}(\mathbf{CD}) = \text{Tr}(\mathbf{DC})$$

$$\text{Tr}(\mathbf{C}+\mathbf{D}) = \text{Tr}(\mathbf{D}) + \text{Tr}(\mathbf{C})$$

$$\mathbf{c}^T\mathbf{c} = \text{Tr}(\mathbf{c}\mathbf{c}^T)$$

Determinant in linear algebra follows different rules as shown below:

$$\det(\mathbf{C}) = \prod_i \lambda_i, \text{ where } \lambda_i = i^{th} \text{ eigenvalue}(\mathbf{C})$$

$$\det(a\mathbf{C}) = a^m \det(\mathbf{C}) \text{ if } \mathbf{C} \in \mathbb{R}^{N \times N}$$

$$\det(\mathbf{C}^T) = \det(\mathbf{C})$$

$$\det(\mathbf{CD}) = \det(\mathbf{C})\det(\mathbf{D})$$

$$\det(\mathbf{C}^{-1}) = 1/\det(\mathbf{C})$$

$$\det(\mathbf{C}^n) = \det(\mathbf{C})^n$$

Machine learning and deep learning algorithms use some derivative matrix in addition to the primary matrix rules. In the following, we explain the derivative rules of matrices:

$$\partial \mathbf{C} = 0 \quad \text{if } \mathbf{C} \text{ is constant}$$

$$\partial(\beta \mathbf{C}) = \beta \partial(\mathbf{C})$$

$$\partial(\mathbf{C} + \mathbf{D}) = \partial(\mathbf{C}) + \partial(\mathbf{D})$$

$$\partial(\mathrm{Tr}(\mathbf{C})) = \mathrm{Tr}(\partial(\mathbf{C}))$$

$$\partial(\mathbf{CD}) = \partial(\mathbf{C})\mathbf{D} + \mathbf{C}\partial(\mathbf{D})$$

$$\partial(\mathbf{C} \circ \mathbf{D}) = \partial(\mathbf{C}) \circ \mathbf{D} + \mathbf{C} \circ \partial(\mathbf{D})$$

$$\partial(\mathbf{C} \otimes \mathbf{D}) = \partial(\mathbf{C}) \otimes \mathbf{D} + \mathbf{C} \otimes \partial(\mathbf{D})$$

$$\partial(\mathbf{C}^{-1}) = -\mathbf{C}^{-1}\partial(\mathbf{C})\mathbf{C}^{-1}$$

$$\partial(\det(\mathbf{C})) = \mathrm{Tr}(\mathrm{adj}(\mathbf{C})\partial\mathbf{C}) = \det(\mathbf{C})\mathrm{Tr}(\mathbf{C}^{-1}\partial\mathbf{C})$$

$$\partial(\ln(\det(\mathbf{C}))) = \mathrm{Tr}(\mathbf{C}^{-1}\partial\mathbf{C})$$

$$\partial\mathbf{C}^{T} = (\partial\mathbf{C})^{T}$$

$$\partial\mathbf{C}^{H} = (\partial\mathbf{C})^{H}$$

In the above equation $\circ$, and $\otimes$ represent element wise and Kronecker product, respectively. Moreover, some necessary derivatives for the matrices, vectors, and scalars can be written as shown below:

$$\frac{\partial \mathbf{c}^{T}\mathbf{s}}{\partial \mathbf{c}} = \frac{\partial \mathbf{s}^{T}\mathbf{c}}{\partial \mathbf{c}} = \mathbf{s}$$

$$\frac{\partial \mathbf{s}^{T}\mathbf{Ch}}{\partial \mathbf{C}} = \mathbf{ah}^{T}$$

$$\frac{\partial \mathbf{s}^{T}\mathbf{C}^{T}\mathbf{h}}{\partial \mathbf{C}} = \mathbf{bh}^{T}$$

$$\frac{\partial \mathbf{s}^{T}\mathbf{Cs}}{\partial \mathbf{C}} = \frac{\partial \mathbf{s}^{T}\mathbf{Cs}}{\partial \mathbf{C}} = \mathbf{ss}^{T}$$

In the equation below, second-order derivatives are shown:

$$\frac{\partial(\mathbf{Sc} + \mathbf{h})^{T}\mathbf{D}(\mathbf{Ec} + \mathbf{k})}{\partial \mathbf{c}} = \mathbf{S}^{T}\mathbf{D}(\mathbf{Ec} + \mathbf{k}) + \mathbf{E}^{T}\mathbf{D}^{T}(\mathbf{Sc} + \mathbf{h})$$

$$\frac{\partial \mathbf{c}^{T}\mathbf{Dc}}{\partial \mathbf{c}} = (\mathbf{D} + \mathbf{D}^{T})\mathbf{c}$$

$$\frac{\partial \mathbf{s}^{T}\mathbf{C}^{T}\mathbf{DCh}}{\partial \mathbf{C}} = \mathbf{D}^{T}\mathbf{Csh}^{T} + \mathbf{DChs}^{T}$$

Consider **C** as a symmetric matrix, then:

$$\frac{\partial}{\partial \mathbf{h}}(\mathbf{a}-\mathbf{Bh})^T \mathbf{C}(\mathbf{a}-\mathbf{Bh}) = -2\mathbf{BC}(\mathbf{a}-\mathbf{Bh})$$

$$\frac{\partial}{\partial \mathbf{a}}(\mathbf{a}-\mathbf{Bh})^T \mathbf{C}(\mathbf{a}-\mathbf{Bh}) = 2\mathbf{C}(\mathbf{a}-\mathbf{Bh})$$

$$\frac{\partial}{\partial \mathbf{B}}(\mathbf{a}-\mathbf{Bh})^T \mathbf{C}(\mathbf{a}-\mathbf{Bh}) = -2\mathbf{C}(\mathbf{a}-\mathbf{Bh})\mathbf{h}^T$$

$$\frac{\partial}{\partial \mathbf{a}}(\mathbf{a}-\mathbf{h})^T \mathbf{C}(\mathbf{a}-\mathbf{h}) = 2\mathbf{C}(\mathbf{a}-\mathbf{h})$$

$$\frac{\partial}{\partial \mathbf{s}}(\mathbf{a}-\mathbf{h})^T \mathbf{C}(\mathbf{a}-\mathbf{h}) = -2\mathbf{C}(\mathbf{a}-\mathbf{h})$$

Furthermore, the following equations represent some rules for the first-order derivative of traces:

$$\frac{\partial}{\partial \mathbf{C}}\text{Tr}(\mathbf{C}) = \mathbf{I}$$

$$\frac{\partial}{\partial \mathbf{C}}\text{Tr}(\mathbf{CD}) = \mathbf{D}^T$$

$$\frac{\partial}{\partial \mathbf{C}}\text{Tr}(\mathbf{C}^T\mathbf{D}) = \frac{\partial}{\partial \mathbf{C}}\text{Tr}(\mathbf{DC}^T) = \mathbf{D}$$

$$\frac{\partial}{\partial \mathbf{C}}\text{Tr}(\mathbf{FCD}) = \mathbf{F}^T\mathbf{D}^T$$

$$\frac{\partial}{\partial \mathbf{C}}\text{Tr}(\mathbf{FC}^T\mathbf{D}) = \mathbf{DF}$$

$$\frac{\partial}{\partial \mathbf{C}}\text{Tr}(\mathbf{D}\otimes\mathbf{C}) = \text{Tr}(\mathbf{D})\mathbf{I}$$

Also, the following equations represent the second-order gradients of traces:

$$\frac{\partial}{\partial \mathbf{C}}\text{Tr}(\mathbf{C}^2) = 2\mathbf{C}^T$$

$$\frac{\partial}{\partial \mathbf{C}}\text{Tr}(\mathbf{C}^2\mathbf{D}) = (\mathbf{CD}+\mathbf{DC})^T$$

$$\frac{\partial}{\partial \mathbf{C}}\text{Tr}(\mathbf{C}^T\mathbf{DC}) = \frac{\partial}{\partial \mathbf{C}}\text{Tr}(\mathbf{DCC}^T) = (\mathbf{D}+\mathbf{D}^T)\mathbf{C}$$

$$\frac{\partial}{\partial \mathbf{C}}\text{Tr}(\mathbf{CDC}^T) = \frac{\partial}{\partial \mathbf{C}}\text{Tr}(\mathbf{DC}^T\mathbf{C}) = \frac{\partial}{\partial \mathbf{C}}\text{Tr}(\mathbf{C}^T\mathbf{CD}) = \mathbf{C}(\mathbf{D}+\mathbf{D}^T)$$

$$\frac{\partial}{\partial \mathbf{C}}\text{Tr}(\mathbf{FCDC}) = \mathbf{F}^T\mathbf{C}^T\mathbf{D}^T + \mathbf{D}^T\mathbf{C}^T\mathbf{F}^T$$

$$\frac{\partial}{\partial \mathbf{C}}\text{Tr}(\mathbf{C}^T\mathbf{C}) = \frac{\partial}{\partial \mathbf{C}}\text{Tr}(\mathbf{CC}^T) = 2\mathbf{C}$$

$$\frac{\partial}{\partial \mathbf{C}}\text{Tr}(\mathbf{C}\otimes\mathbf{C}) = \frac{\partial}{\partial \mathbf{C}}\text{Tr}(\mathbf{C})\text{Tr}(\mathbf{C}) = 2\text{Tr}(\mathbf{C})\mathbf{I}$$

Eigenvalues and eigenvectors

In linear algebra, an eigenvector v_i and eigenvalue λ_i for a matrix C are defined as a non-zero vector and a scalar. Note that the multiplication of v_i and λ_i is equal to the Kronecker product between C and corresponding eigenvector as below:

$$\mathbf{Cv}_i = \lambda_i \mathbf{v}_i$$

An eigenvalue is a factor that operates on the corresponding eigenvector to stretch that along the original direction. Therefore, a negative eigenvalue reverses the direction of the eigenvector.

For a symmetric matrix C, the following equations are true:

$$\mathbf{VV}^T = \mathbf{I} \quad (\mathbf{V} \text{ is orthogonal})$$

$$\lambda_i \in \mathbb{R}$$

$$\mathrm{Tr}(\mathbf{C}^p) = \sum_i \lambda_i^p$$

$$\mathrm{eig}(\mathbf{I}+k\mathbf{C})=1+k\lambda_i$$

$$\mathrm{eig}(\mathbf{C}-k\mathbf{I})=\lambda_i - k$$

$$\mathrm{eig}(\mathbf{C}^{-1})=\lambda_i^{-1}$$

For asymmetric, positive matrix C, the following rules are held:

$$\mathrm{eig}(\mathbf{A}^T\mathbf{A}) = \mathrm{eig}(\mathbf{AA}^T) = \mathrm{eig}(\mathbf{A}) \circ \mathrm{eig}(\mathbf{A})$$

In linear algebra, decomposition is an advanced numerical method for machine learning and deep learning algorithms. In the following, we explain some of the most valuable decompositions methods.

Singular value decomposition for any matrix $\mathbf{C} \in \mathbb{R}^{N \times M}$ is a method that decomposes one matrix into three other matrices and can be written as follows:

$$\mathbf{C} = \mathbf{UDV}^T$$

where

$$\mathbf{U} \in \mathbb{R}^{N \times N} = \text{ eigenvectors of } \mathbf{CC}^T$$

$$\mathbf{D} \in \mathbb{R}^{N \times M} = \sqrt{\mathrm{diag}(\mathrm{eig}(\mathbf{CC}^T))}$$

$$\mathbf{V} \in \mathbb{R}^{M \times M} = \text{ eigenvectors of } \mathbf{C}^T\mathbf{C}$$

LU decomposition

LU decomposition is only defined for a square matrix with non-zero leading principal minors, as follows, where L and U are lower triangular matrix and upper triangular matrix, respectively.

C = LU

Another beneficial decomposition method is named Cholesky decomposition. If the square matrix **C** is a symmetric positive definite matrix, the Cholesky decomposition would be:

$$\mathbf{C} = \mathbf{U}^T\mathbf{U} = \mathbf{LL}^T$$

As before, U is a unique upper triangular matrix, and L is a lower triangular matrix.

Statistics and probabilities

Momentums

First momentum or mean vector for random vector $y \in \mathbb{R}^{N \times 1}$ is defined as follows:

$$\mathbf{M}_1 = \boldsymbol{\mu} = [\mu_1, \mu_2, ..., \mu_N] = [<y_1>, <y_2>, ..., <y_N>]$$

While the second momentum, which is also named covariance matrix, is defined as follows:

$$\mathbf{M}_2 = \mathbf{C} = <(\mathbf{y} - \boldsymbol{\mu})(\mathbf{y} - \boldsymbol{\mu})^T>$$

The following equations describe the third and fourth momentum:

$$\mathbf{M}_3 = <(\mathbf{y} - \boldsymbol{\mu})(\mathbf{y} - \boldsymbol{\mu})^T \otimes (\mathbf{y} - \boldsymbol{\mu})^T>$$
$$\mathbf{M}_4 = <(\mathbf{y} - \boldsymbol{\mu})(\mathbf{y} - \boldsymbol{\mu})^T \otimes (\mathbf{y} - \boldsymbol{\mu})^T \otimes (\mathbf{y} - \boldsymbol{\mu})^T>$$

Expectation

Expectation value can be written as follows:

$$E[\mathbf{SYH} + \mathbf{A}] = \mathbf{S}E[\mathbf{Y}]\mathbf{H} + \mathbf{A}$$
$$Var(\mathbf{Sy}) = \mathbf{S}Var(\mathbf{y})\mathbf{S}^T$$
$$Cov(\mathbf{Sy}, \mathbf{Hz}) = \mathbf{S}Cov(\mathbf{y} + \mathbf{z})\mathbf{H}^T$$
$$E[\mathbf{Sy} + \mathbf{c}] = \mathbf{S}\boldsymbol{\mu} + \mathbf{c}$$

where y and z are random vectors, Y is a random matrix and μ is the mean vector for the y.

Multivariate distributions

Cauchy distribution

The Cauchy distribution is a particular type of t-student distribution and can be written as follows:

$$p(\mathbf{y}) = \pi^{-\frac{N}{2}} \frac{\Gamma(\frac{1+N}{2})\det(\mathbf{C})^{\frac{-1}{2}}}{\Gamma(\frac{1}{2})\left(1 + (\mathbf{y} - \mu)^T \mathbf{C}^{-1}(\mathbf{y} - \mu)\right)^{\frac{1+N}{2}}}$$

where $\mathbf{y} \in \mathbb{R}^{N \times 1}$ and $\mu \in \mathbb{R}^{N \times 1}$ are random vector and mean vector, respectively. $\mathbf{C} \in \mathbb{R}^{N \times N}$ indicates a positive definite covariance matrix and Γ represents the gamma functions.

Dirichlet distribution

Let $\mathbf{y} \in \mathbb{R}^{N \times 1}$ to be a random vector, then the Dirichlet distribution, which is also named multivariate beta distribution (MBD), is defined as follows:

$$p(\mathbf{y}) = \frac{\Gamma(\sum_{i=1}^{N} \lambda_i)}{\prod_{i=1}^{N} \Gamma(\lambda_i)} \prod_{i=1}^{N} y_i^{\lambda_i - 1}$$

where $\mathbf{y} \in \mathbb{R}^{N \times 1}$ is a random vector.

Multimodal distribution

The discrete multimodal distribution $\mathbf{k}$ is defined when the distribution has several modes, which can be shown as distinct peaks in the probability density function. The multivariate probability density function of a multimodal distribution can be written as follows:

$$p(\mathbf{k}) = \frac{k!}{k_1! \cdots k_N!} \prod_{i=1}^{N} p_i^{k_i}, \quad \sum_{i=1}^{N} k_i = N$$

where $\mathbf{k} \in W^{D \times 1}$ is a random vector which consists of whole numbers ($k_i \in 0,1,2,\ldots$) and p_i is the probability of i'th element, hence $0 \leq p_i \leq 1$ and $\sum_{i=1}^{N} p_i = 1$.

Student's t distribution

The multivariate probability density function of a student-t can be shown as follows:

$$p(\mathbf{y}) = (\pi s)^{-\frac{s+N}{2}} \frac{\Gamma(\frac{s+N}{2})\det(\mathbf{C})^{-\frac{1}{2}}}{\Gamma(\frac{s}{2})\left(1 + s^{-1}(\mathbf{y}-\boldsymbol{\mu})^T \mathbf{C}^{-1}(\mathbf{y}-\boldsymbol{\mu})\right)^{\frac{(s+N)}{2}}}$$

where y, μ and s are random vectors, mean vectors, and degree of freedom, respectively, indicate the gamma function. Furthermore, C represents the covariance matrix in the above equation. Note that the student's t distribution equals to Cauchy distribution when s = 1.

Gaussian distribution

The multivariate Gaussian distribution is the most popular form of distribution because its concept is utilized in various applications such as machine learning. Let y be a random vector, then the Gaussian distribution of $\mathbf{y}$ is represented by $\mathbf{y} \sim \mathcal{N}(\boldsymbol{\mu}, \mathbf{C})$ and the probability density function of $\mathbf{y}$ can be written as follows

$$p(x) = \frac{1}{\sqrt{\det(2\pi\mathbf{C})}} \exp\left[-\frac{1}{2}(\mathbf{y}-\boldsymbol{\mu})^T \mathbf{C}(\mathbf{y}-\boldsymbol{\mu})\right]$$

where μ and C are mean vector and covariance matrix, respectively.

CHAPTER 3
Machine Learning

|||||||||||||||||||||||||||||||||||||||||||||||||||||||||||||||||||||||||||||||||||||||||||||||||||||||||||||||||||||||||||||||||||||||||||||||||||||||||||||||||||||||||

Machine learning (ML), as a part of Artificial Intelligence, is the study of computer algorithms that improve automatically through experience and by the use of data. To make predictions or decisions without being programmed to, ML algorithms build a model based on sample data, known as "training data". For extracting information from data, it is hard and not practical to use conventional algorithms. Machine learning methods have been very successful for a wide range of applications in this area. A recent exciting development is the conception of machine learning in the natural sciences. The main goal in this science is to obtain novel scientific insights and findings from observed or simulated data [13]. In statistics, the inputs are often called the predictors, a term we will use interchangeably within formations, and the more classic, all 'y' as the independent variables. In pattern recognition literature [10], the term "features" is preferred. The outputs are called the responses, or classically the dependent variables.

There is a subset of machine learning that has a close relation to computational statistics. Computational statistics focuses on making predictions using computers, but not all machine learning methods are statistical learning. There are methods, theories, and applications in machine learning that the study of mathematical optimization delivers from them. Data mining is one of the studies focusing on exploratory data analysis through unsupervised learning [11]. In its area and business problems, machine learning is applied as predictive analytics.

Machine learning includes computers deciding how to perform their tasks without being explicitly programmed to do so. For this purpose, computers learn from data provided to them, so they carry out specific tasks. For simple tasks assigned to computers, it is possible to manually program algorithms telling the machine how to execute all steps required to solve the problem. Therefore, no learning is needed on the computer's part. In practice, manually programming complex tasks can be challenging; hence it can turn out to be more effective to teach the machine algorithm rather than manually programming every needed step.

The discipline of machine learning employs a wide variety of methods to teach computers to perform tasks where no entirely satisfactory algorithm is available. One approach is to label some of the correct answers as valid to determine correct answers where several potential answers exist. This labeled data can be used as training data to improve its algorithm(s). For example, the MNIST dataset of hand-written digits is a helpful training dataset.

Machine learning approaches

Machine learning approaches based on the nature of the "signal" or "feedback" can be divided into four broad categories:

- Supervised learning: First, the "teacher" introduces example inputs, and their desired outputs which are called labeled data to the system. Therefore, the computer is expected to learn a general rule that maps inputs to outputs [7, 9].
- Semi-supervised learning: Its name represents semi-supervised learning that falls between unsupervised learning (without any labeled training data) and supervised learning (with only labeled training data). Therefore, a few labeled data points with many unlabeled data points [14].
- Unsupervised learning: In this approach, unlabeled data are given to the learning algorithm. So, the computer is expected to learn a general rule on its own. Unsupervised learning can be used to discover hidden patterns in data or as a means towards an end such as what happens in feature learning [15].
- Reinforcement learning: In this approach, the computer program interacts with a dynamic environment that provides feedback that's analogous to rewards. Therefore, the computer system tries to maximize these rewards through navigation into problem space [16].

Other approaches don't fit neatly into this four-category, and sometimes it is better to use more than one approach in one machine learning system.

Historical background

Arthur Samuel, an American at IBM and pioneer in computer gaming and artificial intelligence [18], coined the term *machine learning* in 1959. Nilsson's book on learning machines that mainly deal with machines' capability for pattern classification [19], was a representative book of machine learning research during the 1960s. The popularity of pattern recognition only lasted one decade until the 1970s, as described by Duda and Hart in 1973. In 1981 a report was given on using teaching strategies so that a neural network learns to recognize 40 characters (26 letters, 10 digits, and 4 special symbols) from a computer terminal [21].

Tom M. Mitchell provided a formal definition of the process of learning in the machine learning field: "A computer program is said to learn from experience E concerning some class of tasks T and performance measure P if its performance at tasks in T, as measured by P, improves with experience E" [8]. This definition of the tasks in which machine learning is concerned offers a fundamentally operational definition rather than defining the field in cognitive terms. This follows Alan Turing's proposal in his paper "Computing Machinery and Intelligence," in which the question "Can machines think?" is replaced with the question "Can machines do what we (as thinking entities) can do?" [22].

Modern machine learning has three goals. The first purpose is to solve a classification problem through model development; the second goal is to make predictions for future outcomes based on these models. In cancerous moles classification, a hypothetical algorithm could train the computer vision of moles coupled with supervised learning. The other goal of machine learning is to solve a regression problem and predict a continuous value based on input features. For example, a machine learning algorithm for stock trading that informs future potential predictions is a regression model. Thus, the final purpose of machine learning falls into these three main categories: (a) classification, (b) regression, (c) future prediction.

Data mining

Data mining refers to the process of discovering patterns in large data sets, including methods at the intersection of machine learning, statistics, and database systems [23]. Data mining combines computer science and statistics with an overall goal to intelligently extract information from a data set and transform this information into an understandable structure for further use [24].

Machine learning and data mining overlap significantly because of their methods, but their main focus is different. Machine learning focuses on prediction based on known attributes learned from the training dataset. The main focus in data mining is to discover the (previously) unknown features in the data (this is the analysis step of knowledge discovery in databases). Data mining and machine learning employ other methods but with different purposes. For example, machine learning uses data mining methods for pre-processing steps or to teach an unsupervised algorithm. As said before, the main difference between machine learning and data mining is their basic assumptions. While in machine learning, performance is usually determined based on the ability to reproduce known knowledge, in data mining, the critical task is discovering previously unknown knowledge. A supervised algorithm can work better in machine learning than an unsupervised algorithm because of the valid labeled data. On the other hand, in a typical knowledge discovery or data mining in databases task, supervised methods cannot be used due to the absence of training data.

Optimization

The goal of mathematical optimization (alternatively spelled optimization) or mathematical programming is to choose the best element, based on some standards, from some set of available alternatives. According to optimization problems in all quantitative disciplines such as computer science, engineering, operations research, and economics, finding a mathematical solution has been considered for centuries. First, we must quantify the system's performance using a single number called objective, which could be profit, time, potential energy, or any quantity or combination of quantities shown by a single number [25]. The primary system defines the objective attributes called variables or unknowns. These variables may be restricted or constrained because of the nature of the task. For example, the electron density in a molecule and the interest rate on a loan are always positive. Our goal in optimization is to specify the values of the variables that optimize the objective. Modeling that refers to the whole process of identifying objectives, variables, and constraints for a specific task is the first step of optimization. An appropriate model has the needed complexity for the problem. Simplicity in the model construction may lead to a lack of insights into the practical problem, while the model complexity may be made too difficult to solve. After model construction, an optimization algorithm can be used to find its solution. A computer is usually needed to implement this process due to the complexity of the algorithm and model. Instead of a universal optimization algorithm, there are several algorithms, each of which is suitable for a particular type of optimization problem. Algorithm selection depends on the task that the user tends to perform. This choice is an important one because the problem-solving speed and whether the solution is found depends on this selection.

Machine learning is wholly related to optimization. For example, in many learning problems, minimization of loss function on the training data is essential. The loss function is responsible for identifying the difference between the predicted output and the actual output (for example, in classification problems, the ability of the trained model to correctly allocate a label to instances will compare with the true labels of the data set) [26].

One of the key challenges especially in deep learning is how to find the best weights or parameters. To find the best weights and biases of, for example, a neural network, one of the best approaches is to use an optimizer so that it minimizes the desired cost function. In general, optimizers are algorithms or methods that are used to change the attributes of the neural network iteratively such as the weights and bias terms in order to reduce the losses. To optimize a cost function many questions may arise. For example, how someone should change for instance the weights of a neural network to reduce the losses which are defined by the cost function? Optimization algorithms are responsible for reducing the losses of the model during the training phase and provide the most accurate results possible. There are many types of optimization algorithms which are introduced for training a model in the deep learning area. The most important optimization algorithm is gradient descent. Most of the other algorithms are improved by the gradient descent. Therefore, you should know the gradient descent at first. In this section, we will introduce some important optimization algorithms which are used in deep learning..

• Gradient descent

Gradient Descent is the most basic but most used optimization algorithm. It's used heavily in linear regression and classification algorithms. Backpropagation in neural networks also uses a gradient descent algorithm. Gradient descent is an algorithm to minimize an objective function $J(\theta)$ where the θ is vector of the parameters of the model which are $\theta \in R^d$. The gradient descent updates the parameters in the opposite direction of the gradient of the objective function (which is $\nabla_\theta J(\theta)$) respect to the vector parameters θ. In the update phase of gradient descent, the updating step of parameters can be controlled. The learning rate α determines the size of the steps the gradient descent takes to reach a local or global minimum. In other words, in gradient descent the direction of the slope of the surface created by the objective function follows downhill until it reaches a valley.

Gradient descent is a first order optimization algorithm which is dependent on the first order derivative of the cost function. It calculates which way the weights should be altered so that the function can be reached to a minimum. Through back propagation, the loss is transferred from one layer to another and the parameters of the model (or weights and biases) are modified depending on the losses so that the loss can be minimized. The general updating equation in gradient descent can be written as the following.

$$\theta = \theta - \alpha \nabla_\theta J(\theta) \tag{1}$$

where in the above equation θ is the vector of parameters (for examples vector of weights and biases), $\nabla_\theta J(\theta)$ is the partial derivative of the cost function $J(\theta)$ respect to the θ, and α is the learning rate. While the parameters are updated in the opposite direction of the gradients, you can control the size of updating by the learning rate. The above equation repeats until they reach the minimum of the cost function. In the gradient descent optimization algorithm, in each iteration the gradient of the whole train dataset should compute. Therefore, it is also called batch gradient descent. The gradient descent is guaranteed to converge to the global minimum for convex error surfaces and to a local minimum for non-convex surfaces. For example, the cost function of linear regression is a convex surface and therefore using the gradient descent, you can get the global minimum. However, the cost function for neural networks are non-convex and using the gradient descent, you can reach a local minimum.

The gradient descent optimization algorithm has some advantages and disadvantages. The gradient descent is easy to compute and only the gradient of cost function with respect to parameters is the challenging part. While this optimization problem is easy to compute, it therefore is easy to implement and also easy to understand. However, there are some disadvantages using the gradient descent optimization algorithm. It is sensitive to the weight initialization and it is very probable to tap into the local minima. The other disadvantage of the gradient descent is that for updating weights the gradient on the whole dataset should be able to calculate it. Therefore, if the dataset is too large like an ImageNet dataset, this may take very long time converge to the minima. Also, to compute the gradient for each iteration, you need a high resource which sometimes may not be feasible. In other words, it requires large memory to calculate the gradient on the whole dataset.

• Stochastic gradient descent

Stochastic gradient descent (or SGD) in contrast performs a parameter update for each training example $x^{(i)}$ and label $y^{(i)}$. Therefore, the gradient descent equation is changed to the following equation

$$\theta = \theta - \alpha \nabla_\theta J(\theta; x^{(i)} \cdot y^{(i)}) \tag{2}$$

The gradient descent (or batch gradient descent) performs redundant computations for large datasets, as it recalculates gradients for similar examples before each parameter update. Stochastic gradient descent does away with this redundancy by performing one update at a time. It is therefore usually much faster and can also be used to learn online. Also, stochastic gradient descent performs frequent updates with a high variance that cause the objective function to fluctuate heavily. While gradient descent converges to the minimum, the stochastic gradient descent fluctuation enables it to jump to new and potentially better local minima. On the other hand, this ultimately complicates convergence to the exact minimum, as stochastic gradient descent will keep overshooting. However, it has been shown that when slowly the learning rate is decreased, SGD shows the same convergence behavior as batch gradient descent, almost certainly converging to a local or the global minimum for non-convex and convex optimization respectively. In a nutshell, the stochastic gradient descent has some advantages and disadvantages compared to the gradient descent. The stochastic gradient descent updates the model parameters for each sample. The update per sample in some literatures is called iteration while when all samples update the model at one time called an epoch. Therefore, there are many iterations in each epoch of stochastic gradient descent and therefore, the model updates many times during one epoch. For example, if the training dataset has 10 samples, then each epoch contains 10 iterations. This can imply that the model converges faster and in less time. The other advantage of the stochastic gradient descent is that it requires less memory. This optimization algorithm also may get new local minima which can be better than the gradient descent. In spite of all advantages of stochastic gradient descent, there are some drawbacks. This optimization algorithm has a high variance in model parameters and therefore, may shoot even after achieving global minima. Also, to get the same convergence as gradient descent it needs to slowly reduce the value of learning rate.

• Mini batch gradient descent

The other type of gradient descent optimization algorithm is mini batch gradient descent. It is an improvement on both SGD and standard gradient descent. The mini batch gradient descent takes the advantages of gradient descent and stochastic gradient descent and performs an update for every mini batch of n training examples. The update equation of mini batch gradient descent can be written as the following.

$$\theta = \theta - \alpha \nabla_\theta J(\theta; x^{(i:i+n)} . y^{(i:i+n)}) \tag{3}$$

In the above equation, the gradient of n training examples which referred to the batch respect to the vector parameters is calculated. Like the stochastic gradient descent, the update per sample in some literatures is called iteration while when all samples update the model at one time it is called epoch. Therefore, there are many iterations in each epoch of the stochastic gradient descent and therefore, the model updates many times during one epoch. For example, if the training dataset has 10 samples and the batch size is 2, then each epoch contains 5 iterations. The mini batch gradient descent reduces the variance of the parameter updates which the stochastic gradient descent suffers from. This can lead to a more stable convergence. Also, the mini batch gradient descent can make use of the highly optimized matrix optimizations common to state of the art deep learning libraries that make computing the gradient with respect to a mini batch very efficient. Common mini batch sizes range between 50 and 256, but can vary for different applications. It is best to consider the mini batch size as a power of 2 for better efficiency. Mini batch gradient descent is typically the algorithm of choice when training a neural network. It is easy to understand that if the mini batch size equal to

1, the mini batch gradient descent algorithm converts to stochastic gradient descent. In a nutshell, we can summarize the advantages of the mini batch gradient descent. The first advantage is that this optimization algorithm frequently updates the model parameters and also has less variance. Also, this algorithm requires a medium amount of memory. The mini batch gradient descent, requires less amount of memory compared to gradient descent while it requires more amount of memory compared to stochastic gradient descent. Although, the mini batch gradient descent has some drawbacks. The choosing an optimum value of the learning rate is a challenging and tedious task in mini batch gradient descent and also all types of gradient descent. For example, if the learning rate is too small than gradient descent may take ages to converge while in contrast the gradient descent may not converge. Also, the mini batch gradient descent like the other types may get trapped at local minima.

The gradient descent types of algorithms have challenges which are listed in the following. The first one is how to choose a suitable learning rate. Choosing a proper learning rate can be difficult. A learning rate that is too small leads to painfully slow convergence, while a learning rate that is too large can hinder convergence and cause the cost function to fluctuate around the minimum or even to diverge. Learning rate schedules try to adjust the learning rate during training by reducing the learning rate according to a predefined schedule or when the change in objective between epochs falls below a threshold. These schedules and thresholds have to be defined in advance and are thus unable to adapt to the characteristics of the dataset. We will discuss in more detail about the scheduling of the learning rate. Additionally, the same learning rate applies to all parameter updates. If the data is sparse and the features have very different frequencies, you may not want to update all of them to the same extent, but perform a larger update for rarely occurring features. Another key challenge of minimizing highly non-convex cost functions common for neural networks is avoiding getting trapped in their numerous suboptimal local minima. It is shown that this difficulty arises in fact not from local minima but from saddle points, the points where one dimension slopes up and another slopes down. These saddle points are usually surrounded by a plateau of the same error, which makes it notoriously hard for stochastic gradient descent to escape, as the gradient is close to zero in all dimensions.

• Momentum gradient descent

Stochastic gradient descent and mini batch gradient descent have trouble navigating ravines, which are the areas where the surface curves much more steeply in one dimension than in another. These ravines are usually around local optima. In these scenarios, the stochastic gradient descent and also mini batch gradient descent oscillates across the slopes of the ravine while only making hesitant progress along the bottom towards the local optimum.

Momentum is a method that helps accelerate stochastic gradient descent and mini batch gradient descent in the relevant direction and dampens oscillations. It does this by adding a fraction γ of the update vector of the past time step to the current update vector. The equation of the momentum gradient descent can be written as the following

$$v_t = \gamma v_{t-1} + \alpha \nabla_\theta J(\theta) \tag{4}$$

$$\theta = \theta - v_t \tag{5}$$

The momentum term γ is usually set to 0.9 or a similar value. Using the first equation, the gradients are averaged over a period which has relation to γ. This can dampen the oscillation of the stochastic gradient descent and mini batch gradient descent. Essentially, when using momentum, a ball is pushed down a hill. The ball accumulates momentum as it rolls downhill, becoming faster and faster on the way until it reaches its terminal velocity if there is air resistance. The same thing happens when the parameter updates. The momentum term increases for dimensions whose gradients point in the same directions and reduces updates for dimensions whose gradients change directions. As a result, the momentum gradient descent convergence faster and has less oscillation.

Like the other optimization algorithms, the momentum gradient descent has some advantages and disadvantages. The momentum gradient descent reduces the oscillation of the mini batch gradient descent and also the high variance of the parameters. This causes the momentum gradient descent to converge faster than gradient descent. However, the momentum gradient descent adds one more hyper-parameter which needs to be selected manually and accurately.

• Nesterov accelerated gradient

Momentum may be a good method but if the momentum is too high the algorithm may miss the local minima and may continue to rise up. For example, if a ball that rolls down a hill, blindly following the slope which is unsatisfactory for the optimization problem because it may pass the local minima. If the optimization algorithm has a notion of where it is going so that it knows to slow down before the hill slopes up again, the problem of momentum gradient can be solved. The Nesterov accelerated gradient is introduced to give the momentum gradient descent this kind of prescience. In the momentum gradient descent, the momentum term which is γv_{t-1} is used to move the parameters θ. Therefore, computing the $\theta - \gamma v_{t-1}$ thus gives us an approximation of the next position of the parameters, a rough idea that show where the parameters are going to be. In the Nesterov accelerated gradient, the gradient are calculated respect to the approximate future position of the parameters θ not the respect to the current parameters θ. The updating equation of the Nesterov accelerated gradient can be written as the following equation

$$v_t = \gamma v_{t-1} + \alpha \nabla_\theta J(\theta - \gamma v_{t-1}) \tag{6}$$

$$\theta = \theta - v_t \tag{7}$$

Like the momentum gradient descent, the momentum term in the above equation is usually set to a value of around 0.9. While Momentum first computes the current gradient and then takes a big jump in the direction of the updated accumulated gradient, the Nesterov accelerated gradient first makes a big jump in the direction of the previous accumulated gradient, measures the gradient and then makes a correction, which results in the complete Nesterov accelerated gradient update. This anticipatory update prevents the momentum gradient descent from going too fast and results in increased responsiveness, which has significantly increased the performance of the recurrent neural networks on a number of tasks.

The Nesterov accelerated gradient has some advantages. The first advantage of this optimization algorithm is that this algorithm does not pass the local minima. Also, when the optimization algorithm reaches the local minima, it speeds down while in the momentum gradient descent this does not happen. Although, still, the Nesterov accelerated gradient has one hyper-parameter that should be set manually.

• Adagrad

One of the disadvantages of all the previous optimization algorithms are that the learning rate is constant for all parameters and for each cycle. The Adagrad optimization algorithm changes the learning rate by itself. In other words, the Adagrad is a gradient descent optimization algorithm which adapts the learning rate to the parameters, and performing smaller updates by setting the learning rate small for parameters associated with frequently occurring features, and larger updates by setting the learning rate larger for parameters associated with infrequent features. For this reason, it is well suited for dealing with sparse data. It is shown that the Adagrad can improve the robustness of the mini-batch gradient descent greatly, and therefore it can be used to training large deep networks. Previously, the parameters θ were updated at once and also every parameter θ_i used the same learning rate α. However, the Adagrad algorithm uses a different learning rate for every parameter θ_i at every time step t. Notice that the time step can also interpret as the epoch. For brevity, the g_t is denoted the gradient at time step t while the $g_{t,i}$ is denoted the partial derivative of the objective function respect to the parameter θ_i at time step t. This statement can be written as the following equation:

$$g_{t,i} = \nabla_\theta J(\theta_t, i) \tag{8}$$

The mini batch gradient descent updating equation for every parameter θ_i at each time step t is then can be written as the following:

$$\theta_{t+1,i} = \theta_{t,i} - \alpha g_{t,i} \tag{9}$$

The Adagrad algorithm modifies the learning rate in the update rule at each time step t for every parameter θ_i based on the past gradients that have been computed for θ_i.

$$\theta_{t+1,i} = \theta_{t,i} - \frac{\alpha}{\sqrt{G_{t,ii} + \epsilon}} g_{t,i} \tag{10}$$

where the $G_t \in R^{d \times d}$ is a diagonal matrix where each diagonal element i,i is the sum of the squares of the gradients with respect to θ_i up to time step t, while the ϵ is a smoothing term that avoids division by zero and usually has the order of $1e - 8$. Interestingly, without the square root operation, the algorithm's performance is much worse. As G_t contains the sum of the squares of the past gradients with respect to all parameters θ along its diagonal, the Adagrad can implement and vectorize by performing a matrix and vector product $\odot$ between G_t and g_t. The vectorize version of the Adagrad optimization algorithm can be written as the following

$$\theta_{t+1} = \theta_t - \frac{\alpha}{\sqrt{G_t + \epsilon}} g_t \tag{11}$$

One of the main advantages of Adagrad algorithm is that it eliminates the need to manually tune the learning rate. Most implementations use a default value of 0.01 as a constant value. On the other hand, one of the main weakness of Adagrad is its accumulation of the squared gradients in the denominator. Since every added term is positive, the accumulated sum keeps growing during training. This in turn causes the learning rate to shrink and eventually become infinitesimally small, at which point the algorithm is no longer able to acquire additional knowledge. To solve this problem, the Adadelta optimization algorithm is introduced. The other advantage of the Adagrad algorithm is that the learning rate changes for each parameter. Also, this optimization algorithm can be used for sparse data. Although, this optimization algorithm is computationally expensive because there is a need to calculate the second order derivative. Also, the learning rate is always decreasing results within slow training.

• RMSprop

RMSProp optimization algorithm, is an extension of gradient descent and the Adagrad optimization algorithms that uses a decaying average of partial gradients in the adaptation of the step size for each parameter. The use of a decaying moving average allows the algorithm to forget early gradients and focus on the most recently observed partial gradients seen during the progress of the search, and therefore overcoming the limitation of Adagrad. The RMSprop algorithm is an unpublished, adaptive learning rate method proposed by Geoff Hinton a famous researcher in the deep learning field. The RMSprop and Adadelta have both been developed independently around the same time stemming from the need to resolve the radically diminishing learning rates Adagrad optimization algorithm. The central idea of the RMSprop algorithm is to keep the moving average of the squared gradients for each weight, and then divides the gradient by square root and the mean square. The updating equation of the RMSprop optimization algorithm can be written as the following equation:

$$E[g^2]_t = \beta E[g^2]_{t-1} + (1 - \beta)g_t^2 \tag{12}$$

$$\theta_{t+1} = \theta_t - \frac{\alpha}{\sqrt{E[g^2]_t + \epsilon}} g_t \tag{13}$$

In the above equation ϵ is a smoothing term that avoids division by zero and usually has the order of $1e-8$, β, is hyperparameter of the moving average equation which is usually considered as 0.9. RMSprop optimization algorithm like the Adagrad algorithm divides the learning rate by an exponentially decaying average of squared gradients. Although the RMSprop has improved the Adagrad algorithm, it has two hyperparameters. In the other hand, this optimization algorithm is complex compared to the mini batch gradient descent.

• Adam

Adaptive Moment Estimation (Adam) is another method that computes adaptive learning rates for each parameter. Adam optimization algorithm works with momentums of the first and second order. The intuition behind the Adam is that it does not roll so fast because the optimization algorithm can jump over the minimum, also, there is a need to decrease the velocity a little bit for a careful search. In addition to storing an exponentially decaying average of past squared gradients like Adadelta, Adam also keeps an exponentially decaying average of past gradients. In addition to storing an exponentially decaying average of past squared gradients v_t like Adadelat and RMSprop, Adam algorithm also keeps an exponentially decaying average of past gradients m_t similar to momentum gradient descent. The momentum gradient descent can be interpreted as a ball running down a slope, while the Adam algorithm behaves like a heavy ball with friction, which thus prefers flat minima in the error surface. The decaying averages of past and past squared gradients m_t and v_t can be computed respectively by the following equations.

$$m_t = \beta_1 \, m_{t-1} + (1 - \beta_1)g_t \tag{14}$$

$$v_t = \beta_2 \, v_{t-1} + (1 - \beta_2)g_t^2 \tag{15}$$

where m_t and v_t are estimates of the first moment and the second moment of the gradients respectively. Also, the β_1 and β_2 are the decay rates for the first moment and second moment of gradients respectively. The m_t and v_t are initialized as zero vectors, they are biased towards zero, during the initial time steps, and especially when the decay rates are small (β_1 and β_2 are close to one). These biases can be overcome by computing the bias corrected in the first and second moment estimates as the following:

$$\overline{m}_t = \frac{m_t}{1 - \beta_1^t} \tag{16}$$

$$\overline{v}_t = \frac{v_t}{1 - \beta_2^t} \tag{17}$$

Therefore, the estimation of first and second moments can be used to update the parameters just as similar to the RMSprop optimization algorithm. The Adam update role is written as follows:

$$\theta_{t+1} = \theta_t - \frac{\alpha}{\sqrt{\overline{v}_t} + \epsilon} \overline{m}_t \tag{18}$$

In the above equation ϵ is a smoothing term that avoids division by zero and usually has the order of $1e - 8$. The Adam optimization algorithm has two hyperparameter which should be set for the training. The β_1 and β_2 are usually set 0.9 and 0.999. It is shown that Adam algorithm works well in practice and compares favorably to other adaptive learning method algorithms. The main advantage of the Adam optimization algorithm is that it is too fast and converges rapidly, but also it is a computationally expensive optimization algorithm.

You are now familiar with most of the state of the art optimization algorithms. In a nutshell, Adam is the best optimizer. If one wants to train the neural network in less time and more efficiently it is best practice to use the Adam optimization algorithm.

Statistics

Machine learning and statistics have the same methods, but their principal goal is distinct. While machine learning tries to discover a hidden pattern in data, statistics draw population inferences from a sample [27]. From methodological principles to theoretical tools, the ideas of machine learning have had a long pre-history in statistics. Leo Breiman discriminates data models and algorithmic model paradigms in statistical approaches. Wherein "algorithmic model" [28] refers to the machine learning algorithms like Random statisticians make the combined field statistical learning through adopted methods from machine learning [7, 29].

Theory

A learner is expected to generalize from its experience [30]. Generalization means to perform accurately on new, unseen examples/tasks after having experienced a learning data set. The training data set comes from some generally unknown probability distributions to represent the space of occurrences. The learner could produce sufficiently accurate predictions in new cases through generalization.

Computational learning theory is a branch of theoretical computer science that focuses on the computational analysis of machine learning algorithms and their performance. In learning theory, probabilistic bounds on the performance of algorithms are pretty standard because of the uncertainly of the future and the limited training data sets. For example, the generalization error could quantify using the bias-variance decomposition.

Matching between the complexity of the hypothesis and the complexity of the data's function is one of the critical factors to achieve the best performance. If the hypothesis is less complex than the function, then under-fitting would occur for the model. Increasing the complexity of the model in response could decrease the training error. On the other hand, in cases where the hypothesis is too complex, overfitting would occur for the model, and the generalization will be more flawed.

In addition to performance bounds, the time complexity and feasibility of learning are other remarkable concepts for learning theorists. In computational learning theory, a feasible computation can be done in polynomial time. Positive and negative results are two kinds of time complexity results. Positive results represent that a particular class of functions can be learned in polynomial time. In contrast, the negative results refer to specific classes that cannot be learned in polynomial time.

Different kinds of learning algorithms

Machine learning algorithms follow different approaches, including input/output data and the type of task or problems they are intended to solve.

Supervised learning

Supervised learning algorithms employ a mathematical model to perform a specific task on a training data set. Training data contains a set of training examples, while each of them consists of one or more inputs and the desired output and uses as a supervisory signal. In the mathematical model, each training example is shown by an array or vector, also known as a feature vector, so that a matrix could represent the training data. Learning a function through the iterative optimization of an objective function helps supervised learning algorithms in output prediction for new inputs. An algorithm that uses an optimal function and can accurately predict the output for new inputs that were not a part of the training data. Algorithm improvement refers to the situations where the accuracy of the outputs or predictions increases over time; hence the algorithm is learning to perform that task [8].

Supervised learning algorithms contain three different approaches, including active learning, classification, and regression. Outputs in classification algorithms are limited to distinct classes with a definite set of values, while outputs in regression algorithms may be any numerical value within a range. For instance, an email filtering classifier takes an incoming email as input and predicts the class of the output from the available folders. Moreover, similarity learning is an area of supervised machine learning closely related to regression and classification. As its name implies, similarity learning uses a similarity function to determine the similarity or relation between two objects. It has several applications in the machine learning area, including ranking, recommendation systems, visual identity tracking, face verification, and speaker verification.

Unsupervised learning

Unsupervised learning algorithms employ a mathematical model to find a structure within the unlabeled data set containing only inputs. This structure divides data points into some groups or clusters. In these cases, the algorithms learn from test data that has not been labeled, classified, or categorized. Unsupervised learning algorithms measure relations and similarities in the data and perform a task based on the presence or absence of such similarities in new inputs. While unsupervised learning mainly uses density estimation in statistics, such as finding the probability density function, it has other applications, including summarizing and explaining data features [15].

Cluster analysis divides observations into subsets (called clusters) with the same characteristics according to one or more pre-designated criteria. In other words, all the observations in one specific cluster have similar attributes, while observations of different clusters are dissimilar in the intended features.

Assumptions on the structure of the data are dependent on the clustering techniques. For example, similarity metric and evaluation methods like internal compactness define the similarity between members of the same cluster, while separation measures the difference between clusters. Estimated density and graph connectivity-based techniques are the other methods for this purpose [9, 31]. In the following, we will discuss two important unsupervised learning algorithms.

K means clustering

Clustering is an unsupervised machine learning technique. It is the process of division of the dataset into groups in which the members in the same group possess similarities in features. The commonly used clustering algorithms are K Means clustering, Hierarchical clustering, Density based clustering, Model based clustering. The K means clustering is the simplest and most commonly used iterative type unsupervised learning algorithm. It allows us to cluster the data into different groups and is a convenient way to discover the categories of groups in the unlabeled dataset on its own without the need for any training. It is a centroid based algorithm, where each cluster is associated with a centroid. The main aim of this algorithm is to minimize the sum of distances between the data point and their corresponding clusters. The algorithm takes the unlabeled dataset as input, divides the dataset into K number of clusters, and repeats the process until it does not find the best clusters. The value of K should be predetermined in this algorithm. The value of K is the hyperparameter of the K means clustering algorithm. This value can be found using the Elbow method. The k-means clustering algorithm mainly performs two tasks. In the first step, it determines the best value for K center points or centroids by an iterative process. In the second step, the K means clustering algorithm assigns each data point to its closest k-center. Those data points which are near to the particular k-center, create a cluster.

The K means clustering algorithm does the following steps to find the best cluster for each data point. In the first step, the hyperparameter K which indicates the number of centroids should be initialized. In the second, K random points should be selected and considered as the centroids. These K points can be either the points from the dataset or any other point. In the third step, each of the data points in the dataset should be assigned to their closest centroid. In other words, the distance of each data point from the K centroids should be computed. Afterward, this data point belongs to a cluster with a small distance. There are many distance criteria that can be used in this step such as Euclidean distance or Manhattan distance. To select the distance metric, at first you should know your problem and after that test and try some appropriate distance metric. In the fourth step, the K centroids is updated.

For updating the centroids, many methods can be used. One way that you can use to update the centroids is the average over all data points in each cluster. In other words, the average of all data points in each cluster is calculated and considered as the new centroid. In the fifth step, the last three steps are repeated. This means that clusters in each data point, points to new clusters based on the distance between each of them and the updated centroid and finally updates the centroids. This procedure is repeated until there are not any new assignments which occur in the third step. In other words, the member of each cluster does not change.

The main question in K means clustering algorithm is how to choose the value of K which is the number of clusters? The performance of the K means clustering algorithm depends upon highly efficient clusters that it forms. But choosing the optimal number of clusters is a big task. This is due to the fact that we do not have any additional information about the input data. There are some different ways to find the optimal number of clusters, but the most appropriate method to find the number of clusters or value of K is the Elbow method. The Elbow method is one of the most popular ways to find the optimal number of clusters. This method uses the concept of WCSS value which stands for Within Cluster Sum of Squares, which defines the total variations within a cluster. To find the optimal value of clusters, the elbow method you should follow the following step. In the first step, you should execute the K means clustering on a given dataset for different K values. In the second step, calculate the WCSS value for each value of K. To compute the WCSS value, first calculate the sum of squared distance between each data point and its centroid in a cluster. After calculating the WCSS value for each cluster, take the sum of all these values and consider it as a WCSS value for the K means clustering with K centroids. The WCSS gives you some intuition to how much the centroids and data points in each cluster are similar. In the third step, plot a curve between the calculated WCSS values and the number of clusters K. The sharp point of bend or a point of the plot looks like an arm, then that point is considered as the best value of K. If you plot this curve, you can see it is like and elbow. Since the graph shows this sharp bend, which looks like an elbow, it is hence known as the elbow method. The K means clustering algorithm has some advantages compared to other clustering algorithms. The K means clustering is very smooth in terms of interpretation and resolution. The second advantage of K means compared to hierarchical clustering is that for a large number of variables present in the dataset, K means clustering operates quicker. Also, in K means clustering each instance can modify the cluster in updating the state of centroids. The K means clustering reforms compact clusters and can work on unlabeled numerical data. Moreover, it is fast, robust and uncomplicated to understand and yields the best outcomes when datasets are well distinctive from each other. Despite of all the advantages of K means, it has some drawbacks which leads the researchers to invent new clustering algorithms. The K means clustering algorithm demands for the inferred specification of the number of clusters. Also, the K means clustering fails for nonlinear datasets of data and is unable to deal with noisy data and outliers. Moreover, this clustering algorithm is not suitable for clustering a large dataset. The K means clustering also has some limitations. Sometimes, it is not easy to forecast the number of clusters. In other words, finding the best value of K in this clustering algorithm is not an easy task. The second limitation is that the output is highly influenced by the original input, for example, the number of clusters. In unsupervised learning algorithm, there isn't any information about the data and algorithm which should discover the pattern in the data. However, in K means clustering, the value of K should be specified before clustering the data. Therefore, some researchers state that the K means clustering algorithm is not a pure unsupervised learning algorithm [12, 20].

Principal component analysis

Principal Component Analysis (PCA), is a dimensionality reduction method that is often used to reduce the dimensionality of large datasets by transforming a large set of variables into a smaller one that still contains most of the information from the large set. Reducing the number of variables of a data set naturally comes at the expense of accuracy, but the trick in dimensionality reduction is to trade a little accuracy for simplicity. Because smaller datasets are easier to explore and visualize they make analyzing data much easier and faster for machine learning algorithms without the extraneous variables to process. In nutshell, the idea of the principal component analysis method is to reduce the number of variables of a dataset, while preserving as much information as possible.

High dimensionality means that the dataset has a large number of features. The primary problem associated with high dimensionality in the machine learning field is model overfits, which reduces the ability to generalize beyond the examples in the training set. It is also called the curse of dimensionality. In other words, many learning algorithms can work fine in low dimensions while their performance is decreased with high dimension data. The ability to generalize correctly becomes exponentially harder as the dimensionality of the training dataset grows, as the training set covers a dwindling fraction of the input space. Models also become more efficient as the reduced feature set boosts learning rates and diminishes computation costs by removing redundant features. Principal component analysis can also be used to filter noisy datasets, such as image compression. The first principal component expresses the most amount of variance. Each additional component expresses less variance and more noise, so representing the data with a smaller subset of principal components preserves the signal and discards the noise. Theoretically, the principal component analysis is based on the Pearson correlation coefficient framework and inherits similar assumptions. The first assumption is about the sample size. The minimum of observations should be 150 observations. Ideally it should be a 5:1 ratio of observations to features. The second assumption is about the correlation of features. The feature set is correlated, therefore the reduced feature set effectively represents the original data space. The third assumption is about the linearity. All variables exhibit a constant multivariate normal relationship, and principal components are a linear combination of the original features. Also, principal component analysis assumes that there are not any significant outliers in the data such that they can have a disproportionate influence on the results.

The main procedure of the principal component analysis is as follows. In the first step the data should be standardized. The goal of this step is to standardize the range of the continuous initial variables so that each one of them contributes equally to the analysis. More specifically, the reason why it is critical to perform standardization before using the principal component analysis is that the latter is quite sensitive regarding the variances of the initial variables. That is, if there are large differences between the ranges of initial variables, those variables with larger ranges will dominate over those with small ranges, which will lead to biased results. For example, a variable that ranges between 0 and 100 will dominate over a variable that ranges between 0 and 1. Therefore, transforming the data to comparable scales can prevent this problem. Mathematically, this can be done by subtracting the mean and dividing it by the standard deviation for each value of each variable. After the standardization, all the variables will be transformed to the same scale.

$$z = \frac{x - \mu}{\sigma^2} \tag{19}$$

In the second step the covariance matrix should be calculated. In this step, you can understand how the variables of the input data set are varying from the mean with respect to each other. In other words, you can discover if there is any relationship between them. Sometimes, variables are highly correlated in such a way that they contain redundant information. Therefore, in order to identify these correlations, the covariance matrix is calculated. The covariance matrix is a $p \times p$ symmetric matrix that has as entries of the covariance associated with all possible pairs of the initial variables. The value of p comes from the number of dimensions. For example, for a 3 dimensional dataset with 3 variables, the covariance matrix is a 3×3. Since the covariance of a variable with itself is its variance, the variances of each initial variable are placed in the main diagonal of the covariance matrix. Moreover, the covariance is commutative (in other words it is symmetric), the entries of the covariance matrix are symmetric with respect to the main diagonal, which means that the upper and the lower triangular portions are equal. In the covariance matrix, the sign of each covariance is important. If the covariance is positive, we can conclude that the two variables increase or decrease together. In other words, these two variables are correlated if their covariance is positive. In contrast, if this value is negative, that is one increases when the other decreases or simply they are uncorrelated. The third step in principal component analysis is to compute the eigenvectors and eigenvalues of the covariance matrix to identify the principal components. Eigenvectors and eigenvalues are the linear algebra concepts that are necessary to compute from the covariance matrix in order to determine the principal components of the data. Before getting to the explanation of these concepts, let's first understand what the meaning of the principal components is. Principal components are new variables that are constructed as linear combinations or mixtures of the initial variables. These combinations are done in such a way that the new variables are uncorrelated and most of the information within the initial variables is squeezed or compressed into the first components. Therefore, the idea is 10 dimensional data gives you 10 principal components, but the goal of principal component analysis is to put the maximum possible information in the first component, then maximum remaining information in the second and so on. Organizing information in principal components this manner, will allow you to reduce dimensionality without losing much information, and by discarding the components with low information and then considering the remaining components as your new variables. If you keep all of the principal components, it is obvious that you do not lose any information. An important thing to realize here is that, the principal components are less interpretable and do not have any real meaning since they are constructed as linear combinations of the initial variables. Geometrically speaking, the principal components represent the directions of the data that explain a maximal amount of variance, that is to say, the lines that capture most of the information from the data. The relationship between variance and information here, is that, the larger the variance carried by a line, the larger the dispersion of the data points along it, and the larger the dispersion along a line, the more the information it has.

Since there are as many principal components as there are variables in the data, the principal components are constructed in such a manner that the first principal component accounts for the largest possible variance in the data set. The first principal component has the high energy of the input dataset and most of the information of the dataset is in it. The second principal component is calculated in the same way, with the condition that it is uncorrelated with the first principal component and that it accounts for the next highest variance. In other words, the second principal component should be perpendicular to the first principal component. The same procedure is done to calculate the third principal component. Notice that the third principal component is uncorrelated with the first and second principal components. This continues until a total of p principal components have been calculated, equal to the original number of variables.

The key element in principal component analysis is eigenvectors and eigenvalues. This is due to the fact that the eigenvectors of the covariance matrix are actually the directions of the axes where there is the most variance or most information which was called the principal components previously. The eigenvalues are simply the coefficients attached to eigenvectors, which give the amount of variance carried in each principal component. In other words, the eigenvalues can tell us how much the energy of the signal or dataset is in its corresponding eigenvector. By ranking the descending eigenvectors in order of their eigenvalues, you get the principal components in order of significance. The first eigenvalues have the most information while the last ones have the least information about the dataset or signal.

In the fourth step of principal component analysis, the feature vectors are extracted. Computing of the eigenvectors and ordering them by their eigenvalues in descending order, allows us to find the principal components in order of significance. In this step, we choose whether to keep all these components or discard those of lesser significance which are with low eigenvalues, and form with the remaining ones a matrix of vectors that we call the Feature vector. Therefore, the feature vector is a matrix that has as columns of the eigenvectors of the components that we decide to keep. This makes it the first step towards dimensionality reduction, because if we choose to keep only p eigenvectors out of n, the final data set will have only p dimensions. To extract the feature matrix, you can easily set a threshold on the eigenvalues vector and keep the eigenvalues that are greater than the threshold. In the last step, the data should be recast along the principal components axes. In the previous steps, apart from standardization, you do not make any changes on the data and just select the principal components and form the feature vector, but the input dataset remains always in terms of the original axes. In this step, which is the last one, the goal is to use the feature vector formed using the eigenvectors of the covariance matrix, to reorient the data from the original axes to the ones represented by the principal components. This can be done by multiplying the transpose of the original data set by the transpose of the feature vector [20].

Semi-supervised learning

In semi-supervised learning, the algorithm has taken a hint about how to construct the categories. For example, in text document classification, only a tiny portion of the training data is labeled, which means a large part of the training data is not labeled. Here, the trained model is expected to categorize the unlabeled data accurately. In other words, semi-supervised algorithms can learn from partially labeled data sets. Text documents have a wide variety of applications in the machine learning area, such as volumes and volumes of scripts, books, blogs, etc., which are primarily unlabeled. As data labeling is an expensive and time-consuming process, it is reasonable to use semi-supervised techniques.

Semi-supervised learning fell between unsupervised learning (with unlabeled training data) and supervised learning (with only labeled training data), which means a significant part of the training examples are without labels. Many machine-learning researchers have found that this combination of labeled and unlabeled training data can improve the learning capacity of the algorithms. Even the noisy, finite, and imprecise training data of the weakly supervised learning results in larger practical training sets.

Reinforcement learning

The reinforcement learning (RL) algorithm can learn through the environment's reactions. An RL model receives the implication of its actions in the form of reward or punishment. Therefore, it has to choose its next actions based on its past experiences (exploitation) and new choices (exploration), which is the principle of trial-and-error learning. The environment sends a numerical reward to the RL agent. This reinforcement signal encodes the success of an action's outcome, where a successful action increases the cumulative reward. This helps the agent to learn to select actions that maximize the accumulated reward over time (Here, the term reward is used to indicate a neutral mood and does not express any pleasure, hedonic impact, or other psychological states) [32].

In fact, reinforcement learning is an area of machine learning that focuses on reward maximizing through taking correct actions in an environment. Game theory, control theory, operations research, information theory, simulation-based optimization, multi-agent systems, swarm intelligence, statistics, and genetic algorithms are some other disciplines that employ reinforcement learning. The environment in machine learning refers to a Markov decision process (MDP). Reinforcement learning algorithms are effective when exact models are infeasible. Hence most of them do not utilize a fixed mathematical model of the MDP, which means they apply dynamic programming techniques [33]. Reinforcement learning algorithms are used in a self-driving car or in games as a computer opponent.

Deep reinforcement learning is one the hot topics in the last few years. Deep reinforcement learning combines artificial neural networks with a framework of reinforcement learning that helps software agents learn how to reach their goals. That is, it unites function approximation and target optimization, mapping states and actions to the rewards they lead to. While neural networks are responsible for recent AI breakthroughs in problems like computer vision, machine translation and time series prediction—they can also be combined with reinforcement learning algorithms to create something astounding like Deepmind's AlphaGo, an algorithm that beat the world champions of the Go board game. Also, the reinforcement learning algorithms that incorporate deep neural networks can beat human experts playing numerous Atari video games, Starcraft II and Dota-2. That is why the world focuses on deep reinforcement learning. Reinforcement learning refers to goal oriented algorithms, which learn how to achieve a complex objective or goal. In other words, the aim of reinforcement learning is find out how to maximize a cost function over many steps. For example, they can maximize the points won in a game over many moves. Reinforcement learning algorithms can start from a blank slate, and under the right conditions, achieve superhuman performance. Let's take an example and explain details of reinforcement learning in this example. Consider how a child learns to walk. In the first step, the baby should learn how to stand on his or her feet. In the first try, the baby stands and immediately falls to the ground. This hurts the baby and punishes the baby. Therefore, the baby learns from his or her first try and uses this experience in the second try. In the second try, he or she can stand for a few second and the parents encourage the baby. This a reward for the baby. The baby tries based on these rewards and punishments until he or she can walk without falling to the ground. The reinforcement learning basic idea is very similar to how the baby learns to walk. A set of rewards and punishments are defined for the model and the model should find the best weights that maximize the rewards and minimize the punishments. In other words, the reinforcement learning model should maximize or minimize a cost function which these rewards and punishments to be included in it.

Reinforcement learning solves the hard problem of correlating immediate actions with the delayed outcomes they produce. Like humans, reinforcement learning algorithms sometimes have to wait to see the result of their decisions. They operate in a delayed return environment, where it can be difficult to understand which action leads to which outcome over many time steps. Like the baby learns how to walk during the time, the reinforcement learning algorithms are slowly performing better and better in more ambiguous, real world environments while choosing from an arbitrary number of possible actions, rather than from the limited options of a repeatable video game. That is, they are beginning to achieve goals in the real world. DeepMind claimed in May 2021 that reinforcement learning was probably sufficient to achieve artificial general intelligence. Also, companies are starting to apply deep reinforcement learning to problems in industry especially in the robotics industry. For example, Pieter Abbeel's Covariant uses deep reinforcement learning in industrial robotics. Pathmind applies deep reinforcement learning to simulations of industrial operations and supply chains to optimize factories, warehouses and logistics. Google is applying deep reinforcement learning to problems such as robot locomotion and chip design, while Microsoft relies on deep reinforcement learning to power its autonomous control systems technology.

Until now, you are familiar with the concept of reinforcement learning and know it is one of the hottest topics in academia and industry. Now, let's define some basic definitions in reinforcement learning context. Reinforcement learning can be understood through the concepts of agents, environments, states, actions and rewards. In the following, all of these concepts will be defined. Before we start, remember that capital letters tend to denote sets of things, and lowercase letters denote a specific instance of that thing. For example A is all possible actions, while a is a specific action contained in the set.

- Agent: An agent takes actions. For example, a drone making a delivery, Super Mario navigating a video game, or a baby learns to walk. The algorithm is the agent. It may be helpful to consider that in life, the agent is you.

- Action (A): A is the set of all possible moves or actions the agent can make. An action is almost self-explanatory, but it should be noted that agents usually choose from a list of discrete, possible actions. In video games, the list might include running right or left, jumping high or low, crouching or standing still. In the stock markets, the list might include buying, selling or holding any one of an array of securities and their derivatives. When handling aerial drones, alternatives would include many different velocities and accelerations in 3D space. When a baby wants to walk, the action list might include standing and take a step.

- Discount factor: The discount factor is multiplied by future rewards as discovered by the agent in order to dampen effects of these rewards on the agent's choice of action. It is designed to make future rewards worth less than immediate rewards. This enforces a kind of short-term hedonism in the agent. Often expressed with the lowercase Greek letter gamma (γ). If γ is 0.8, and there is a reward of 10 points after 3 time steps, the present value of that reward is equal to $0.8^3 \times 10$. A discount factor of 1 would make future rewards worth just as much as immediate rewards.

- Environment: The world through which the agent moves, and which responds to the agent. The environment takes the current state of the agent and action as input, and returns as the output as the reward of the agent and its next state. If you are the agent, the environment could be the laws of physics and the rules of society that process your actions and determine the consequences of them.

- State (*S*): A state is a concrete and immediate situation in which the agent finds itself, in other words a specific place and moment, an instantaneous configuration that puts the agent in relation to other significant things such as tools, obstacles, enemies or prizes. It can the current situation returned by the environment, or any future situation.

- Reward (*R*): A reward is the feedback by which we measure the success or failure of an agent's actions in a given state. For example, in a video game, when Mario touches a coin, he wins points, or when a baby takes some steps, he or she encounter with Encourage of parents. From any given state, an agent sends output in the form of actions to the environment, and the environment returns the new state of agent which resulted from acting on the previous state as well as rewards. Rewards can be immediate or delayed. They effectively evaluate the agent's action.

- Policy (*π*): The policy is the strategy that the agent employs to determine the next action based on the current state. It maps states to actions, the actions that promise the highest reward.

- Value (*V*): The expected long-term return with a discount, as opposed to the short-term reward *R*. $V\pi(s)$ is defined as the expected long-term return of the current state under policy *π*. You discount rewards, or lower their estimated value, the further into the future they occur.

- Q-value or action-value (*Q*): Q-value is similar to Value, except that it takes an extra parameter, the current action a. $Q\pi(s.a)$ refers to the long-term return of an action taking action *a* under policy *π* from the current state *s*. *Q* maps state-action pairs to rewards. Note the difference between *Q* and policy.

- Trajectory: A sequence of states and actions that influence those states.

- Key distinctions: Reward is an immediate signal that is received in a given state, while value is the sum of all rewards you might anticipate from that state. Value is a long-term expectation, while reward is an immediate pleasure. Value is eating spinach salad for dinner in anticipation of a long and healthy life while reward is eating cocaine for dinner and to hell with it. They differ in their time horizons. Therefore, you can have states where value and reward diverge. However, you might receive a low, immediate reward (like spinach) even as you move to position with great potential for long-term value, or you might receive a high immediate reward (like cocaine) that leads to diminishing prospects over time. This is why the value function, rather than immediate rewards, is what reinforcement learning seeks to predict and control.

So, environments are functions that transform an action taken in the current state into the next state and a reward, agents are functions that transform the new state and provide rewards into the next action. You may know and set the agent's function, but in most situations where it is useful and interesting to apply reinforcement learning, you do not know the function of the environment. It is a black box where we only see the inputs and outputs. It is like most people's relationship with technology. In other words, you know what technology does, but you do not know how it works. Reinforcement learning represents an agent's attempt to approximate the environment's function, such that you can send actions into the black box environment that maximize the rewards it spits out. Reinforcement learning learns from actions by the results they produce. It is goal oriented, and its aim is to learn the sequences of actions that will lead an agent to achieve its goal, or maximize its objective function. For example, in video games, the goal is to finish the game with the most points, therefore each additional point obtained throughout the game will affect the agent's subsequent behavior. In other words, the agent may learn that it should shoot battleships, touch coins or dodge meteors to maximize its score. The other example is that in the real world, the goal might be for a robot to travel from point A to point B, and every inch the robot is able to move closer to point B could be counted like points. There is an objective function in the reinforcement learning algorithm which should maximized. In the following, there is an example of an objective function for reinforcement learning.

$$f = \sum_{t=0}^{\infty} \gamma^t r(x(t).a(t)) \tag{20}$$

where in the above function x is the state at a given time step, and a is the action taken in that state, r is the reward function for x and a over the time t, which stands for time steps. Therefore, this objective function calculates all the rewards that the agent could obtain by running through a game.

The reinforcement learning differs from both supervised learning and unsupervised learning. The reinforcement learning and supervised learning both use a mapping function between input and output, but in supervised learning feedback provided to the agent is a correct set of actions for performing a task, while reinforcement learning uses rewards and punishment as signals for positive and negative behavior. As compared to unsupervised learning, reinforcement learning is different in terms of goals. While the goal in unsupervised learning is to find similarities and differences between data points, in reinforcement learning the goal is to find a suitable action model that would maximize the total cumulative reward of the agent.

The goal of reinforcement learning is to pick the best known action for any given state, which means the actions have to be ranked, and assigned values relative to one another. Since those actions are depending on the state, what the reinforcement learning is really gauging is the value of state-action pairs, that is an action taken from a certain state, something you did somewhere else. Here are a few examples to demonstrate that the value and meaning of an action is contingent upon the state in which it is taken. The first example is that if the action is marrying someone, then marrying a 35 year old person when you are 18 probably means something different than marrying a 35 year old person when you are 90, and these two outcomes probably have different motivations and lead to different outcomes. The other example is that if the action is yelling "Fire!", then performing the action a crowded theater should mean something different from performing the action next to a squad of men with rifles. We cannot predict the outcome of an action without knowing the context. In reinforcement learning the state-action pairs maps to the values that are expected them to produce with the Q function. The Q function takes the state of agent and action as its input, and maps them to probable rewards.

Reinforcement learning is the process of running the agent through sequences of state-action pairs, observing the rewards that result, and adapting the predictions of the Q function to those rewards until it accurately predicts the best path for the agent to take. That prediction is known as a policy. Also, the reinforcement learning is an attempt to model a complex probability distribution of rewards in relation to a very large number of state-action pairs. This is one reason reinforcement learning is paired with a Markov decision process, a method to sample from a complex distribution to infer its properties. It closely resembles the problem that inspired Stan Ulam to invent the Monte Carlo method, trying to infer the chances that a given hand of solitaire will turn out successful. Any statistical approach is essentially a confession of ignorance. The immense complexity of some phenomena (like biological, political, sociological, or related to board games) make it impossible to reason from first principles. The only way to study them is through statistics, measuring superficial events and attempting to establish correlations between them, even when we do not understand the mechanism by which they relate. Reinforcement learning, like deep neural networks, is one such strategy, relying on sampling to extract information from data. After a little time spent employing something like a Markov decision process to approximate the probability distribution of reward over state-action pairs, a reinforcement learning algorithm may tend to repeat actions that lead to reward and cease to test alternatives. There is a tension between the exploitation of known rewards, and continued exploration to discover new actions that also leads to victory. Just as oil companies have the dual function of pumping crude out of known oil fields while drilling for new reserves, so well, reinforcement learning algorithms can be made to both exploit and explore to varying degrees, in order to ensure that they don't pass over rewarding actions at the expense of known winners. As said before, the reinforcement learning algorithm is iterative. In its most interesting applications, it does not begin by knowing which rewards state-action pairs will produce. It learns those relations by running through states again and again, like athletes or musicians iterate through states in an attempt to improve their performance.

You could say that an algorithm is a method to more quickly aggregate the lessons of time. Reinforcement learning algorithms have a different relationship to time than humans do. An algorithm can run through the same states over and over again while experimenting with different actions, until it can infer which actions are best from which states. Effectively, algorithms enjoy their very own 'Groundhog Day', where they start out as dumb jerks and slowly get wise. Since humans never experience 'Groundhog Day' outside of the movie, the reinforcement learning algorithms have the potential to learn more, and better, than humans. Indeed, the true advantage of these algorithms over humans stems not so much from their inherent nature, but from their ability to live in parallel on many chips at once, to train night and day without fatigue, and therefore to learn more. An algorithm trained on the game of Go, such as AlphaGo, will have played many more games of Go than any human could hope to complete in 100 lifetimes.

One interpretation of neural networks is that they estimate mapping function (the function which maps the input and output) or they are function approximators, which are particularly useful in reinforcement learning when the state space or action space are too large to be completely known. A neural network can be used to approximate a value function, or a policy function. That is, neural nets can learn to map states to values, or state-action pairs to Q values. Rather than use a lookup table to store, index and update all possible states and their values, which is impossible with very large problems, one can train a neural network on samples from the state or action space to learn to predict how valuable those are relative to our target in reinforcement learning. Like all neural networks, they use coefficients to approximate the function relating inputs to outputs, and their learning consists to finding the right coefficients, or weights, by iteratively adjusting those weights along gradients that promise less error. In reinforcement learning, convolutional networks can be used to recognize the state of an agent when the input is visual. For example, the screen that Mario is on, or the terrain before a drone. That is, they perform their typical task of image recognition.

But convolutional networks derive different interpretations from images in reinforcement learning than in supervised learning. In supervised learning, the network applies a label to an image; that is, it matches names to pixels. In fact, it will rank the labels that best fit the image in terms of their probabilities. Consider an image of a donkey, this network might decide the picture is 80% likely to be a donkey, 50% likely to be a horse, and 30% likely to be a dog. In reinforcement learning, given an image that represents a state, a convolutional net can rank the actions possible to perform in that state. For example, it might predict that running right will return 5 points, jumping 7, and running left none. Consider the super Mario game that super Mario should decide whether runs or jumps. This example illustrates what a policy agent does, mapping a state to the best action.

$$a = \pi(s) \tag{21}$$

The above equation is a policy that maps a state to an action. If you recall, this is distinct from Q, which maps state action pairs to rewards. To be more specific, Q maps state-action pairs to the highest combination of immediate reward with all future rewards that might be harvested by later actions in the trajectory. Having assigned values to the expected rewards, the Q function simply selects the state-action pair with the highest so-called Q value.

At the beginning of reinforcement learning, the neural network parameters (weights and biases) may be initialized stochastically, or randomly. The initialization of weights of the neural networks is discussed in the previous sections. Using feedback from the environment, the neural net can use the difference between its expected reward and the ground-truth reward to adjust its weights and improve its interpretation of state-action pairs. This feedback loop is analogous to the backpropagation of error in supervised learning. However, supervised learning begins with knowledge of the ground-truth labels the neural network is trying to predict. Its goal is to create a model that maps different images to their respective names. Reinforcement learning relies on the environment to send it a scalar number in response to each new action. The rewards returned by the environment can be varied, delayed or affected by unknown variables, introducing noise to the feedback loop. This leads us to a more complete expression of the Q function, which takes into account both the immediate rewards produced by an action and the delayed rewards that may be returned several time steps deeper in the sequence.

Reinforcement learning is about making sequential decisions to attain a goal over many steps. While other types of artificial intelligence (AI) perform what you might call perceptive tasks, like recognizing the content of an image, reinforcement learning performs tactical and strategic tasks. Games are a good proxy for problems that reinforcement learning can solve, but reinforcement learning is also being applied to real-world processes in the private and public sectors like robotics, industrial operations, supply chain & logistics, traffic control, bidding & advertising recommender systems, load balancing, and augmented natural language processing (NLP).

The process of reinforcement learning is one of the greatest ideas in (AI). But there are challenges when you work on reinforcement learning algorithms. The main challenge in reinforcement learning lays in preparing the simulation environment, which is highly dependent on the task to be performed. For example, when the model has to go superhuman in Chess, Go or Atari games, preparing the simulation environment is relatively simple. When it comes to building a model capable of driving an autonomous car, building a realistic simulator is crucial before letting the car ride on the street. The model has to figure out how to brake or avoid a collision in a safe environment, where sacrificing even a thousand cars comes at a minimal cost. Transferring the model out of the training environment and into the real world is where things get tricky. The second challenge is that scaling and tweaking the neural network controlling the agent is another challenge. There is no way to communicate with the network other than through the system of rewards and penalties. This in particular may lead to catastrophic forgetting, where acquiring new knowledge causes some of the old to be erased from the network. The third challenge is reaching a local optimum which is the agent performs the task as it is, but not in the optimal or required way. A "jumper" jumping like a kangaroo instead of doing the thing that was expected of it-walking-is a great example. The fourth challenge is that there are agents that will optimize the prize without performing the task it was designed to do. An interesting example of this can be found in the OpenAI CoastRunners game, where the agent learned to gain rewards, but not to complete the race. This is due to the bad definition of the reward functions which causing odd reinforcement learning model behavior.

Although reinforcement learning, deep learning, and machine learning are interconnected no one of them in particular is going to replace the others. Yann LeCun, the renowned French scientist and head of research at Facebook, jokes that reinforcement learning is the cherry on a great AI cake with machine learning the cake itself and deep learning the icing. Without the previous iterations, the cherry would top nothing. Also, some researchers believe that the reinforcement learning is the future of machine learning [12, 17].

Self-learning

In the machine learning area, self-learning and its neural network capable named crossbar adaptive array (CAA) was introduced in 1982 [34]. The algorithm is self-learning and is expected to learn without any external rewards or external teacher advice. The CAA self-learning algorithm is responsible for performing two basic steps; first, making decisions about actions, and second, measuring the emotions (feelings) about consequence situations. In fact, the interaction between cognition and emotion is the principal element of these systems [35].

Situation and action (behavior) is the only input and output of the system, respectively. Therefore, a reinforcement signal or an advice input has no role in this method. The emotion toward the consequence situation determines the backpropagated value (secondary reinforcement). Two environments are defined for the CAA; Genetic environment and behavioral environment. The genetic environment sends initial emotions about situations to help the system behavior in a behavioral environment. In this fashion, the CAA receives the genome (species) vector from the genetic environment in order to learn a goal-seeking behavior in an environment that contains both desirable and undesirable situations [36].

Feature learning

Various learning algorithms such as principal components analysis and cluster analysis seek better representations of input data through the training step. Feature learning algorithms, or representation learning algorithms, generally try to keep the input information and also modify it to make it useful. It is usually done as a pre-processing step before classification or predictions. This modification allows the algorithm to reconstruct a set of input data that comes from the unknown data-generating distributions without staying necessarily faithful to settings that are unacceptable under that distribution. After that, there is no need for manual feature engineering, but the machine can also learn the features and perform the desired task. The feature learning system employs a set of techniques to automatically find the hidden pattern of features in raw data and classify them.

Some tasks like classification need to process their input mathematically and computationally. This fact motivated feature learning. Real-world data such as images, video, and sensor data cannot be defined algorithmically with specific features. In these cases, it is possible to find such features or representations through examination instead of relying on explicit algorithms. Feature learning performs as supervised or unsupervised algorithms.

- In supervised feature learning, there are labeled input data that is used for feature learning. Examples of this type of learning are supervised neural networks, multi-layer perceptron, and (supervised) dictionary learning.
- In unsupervised feature learning, there are unlabeled input data that is used for feature learning. Examples of this type of learning contain dictionary learning, independent component analysis [37], autoencoders [38], matrix factorization [39], and different clustering methods [40].

Machine learning models

One of the most important parts of machine learning is to create a model. This model is trained on certain training data and can then process additional data to make predictions. Various types of models have been used and researched for machine learning systems.

Linear regression

Linear regression is a supervised machine learning algorithm. The output of this learning algorithm is continuous and can therefore be classified into the regression category. The goal of linear regression is to predict the value of a variable based on the value of another variable. The output variable or the variable you want to predict is called the dependent variable. The input variable or the variable you are using to predict the output variable value is called the independent variable. The input variable is also called input features. Linear regression uses a linear equation to predict the output variable. The linear equation of linear regression can be written as the following:

$$\tilde{y} = wx + b \tag{22}$$

where w is coefficient vector, x is the input vector for observation, b is the bias term, and y is the output variable or the predicted variable. In linear regression, we want to estimate the coefficients of the linear equation, involving one or more independent variables that best predict the value of the dependent variable. Linear regression fits a straight line or surface that minimizes the discrepancies between predicted and actual output values. To find the best prediction of the output value, you should measure the distance between the prediction and the actual value of the y variable. The simple linear regression uses the least squares method to find out the best-fit line for a set of paired data.

One of the advantages of the linear regression models is that they are relatively simple and provide an easy to interpret mathematical formula that can generate predictions. Linear regression can be applied to various areas in business and academic study like predicting the stock price, house price, and many others. Also, linear regression is used in biological, behavioral, environmental, social sciences, and business. Linear regression models have become a proven way to scientifically and reliably predict the future. Because linear regression is established in statistics, the properties of linear regression algorithms are well understood and can be trained very quickly. In linear regression, it is important to note that the dependent and independent variables should be quantitative.

In linear regression, the input values or features vector x combines so that the predicted output is close enough to its actual output value. As such, both the input values and the output value are numeric. The linear equation assigns one scale factor to each input value or feature, which is called weight. One additional coefficient is also added, giving the line an additional degree of freedom and is often called the intercept or the bias weight. In two dimensional space, this term moves up or down a line. For example, in a simple two dimensional regression problem which is a single input and a single output, the form of the above equation would be

$$\tilde{y} = w_1 x + b \tag{23}$$

In higher dimensions when there is more than one input or feature, the line is called a plane or a hyperplane. Therefore, the weights are the hyperplane parameters and w converts to a vector. The complexity of the linear regression model is defined by the number of weights used in the model or equivalently the number of input features. It is obvious that by increasing the input features or the number of weights, the complexity of the linear model will increase.

By now, you are familiar with the concept of linear regression. Now, we want to talk about how to train a linear regression and test it in practice. Most of the supervised learning algorithms have some unknown parameters which should be estimated from the training dataset. In the linear regression model, the learning process means estimating the values of the weights used in the representation with the train dataset. During the training phase, the linear regression model learns which features are important and which of them does not have significant effects on the output variable. This knowledge about the training dataset, abstracts in the weights of the model. The large value of weight indicates the importance of the feature and if a weight has a low value (close to zero), it means that this feature has a negative effect or does not affect the output value significantly. To train linear regression, we need to define how well the model performs on the training dataset. This need leads us to define a new function which in machine learning is called loss function. The loss function measures how well our model is. There is some loss function for linear regression. The most famous one is the least squares error which is represented by the following equation.

$$L_y = (y - \tilde{y})^2 \tag{24}$$

where the y is the actual output value and $\tilde{y}$ is the prediction of the linear regression model. The above equation measures the loss for one sample. To compute the overall loss on the training dataset, we should compute the loss for all train samples and calculate their average. This new function in some references is called cost function. The linear regression cost function can be written as the following equation.

$$C = \frac{1}{N} \sum_{i=1}^{m} L_{y^{(i)}} = \sum_{i=1}^{m} (y^{(i)} - \tilde{y}^{(i)})^2 \tag{25}$$

where the $L_{y^{(i)}}$ is the loss for the ith sample in the training dataset, $y^{(i)}$ is the actual ith output, the $\tilde{y}^{(i)}$ is the ith linear regression prediction value, and m is the number of training samples. The reason to use the quadratic form of error in the above equation is the derivability of the cost function in all points. The interpretation of the above equation is that given a regression line through the data, the distance from each data point to the regression line should calculate, square it, and sum all of the squared errors together. Now we have a cost function that measures how well the linear regression is on the training dataset, it is time to talk about how to train linear regression. To train and find the best weights is equivalent to finding the best weights which minimize the above cost function. Now the training process of linear regression converts to a minimization problem. There are some approaches to solve this optimization problem. The most popular approach is to use a process of optimizing the values of the coefficients by iteratively minimizing the error of the model on the training dataset. This operation is called gradient descent and works by starting with random values for each coefficient. By using gradient descent, the weights of linear regression update in some iteration. The main idea of the gradient descent is to move in the opposite direction of the partial derivate of each weight with respect to the cost function. In other words, the weights are updated in the direction towards minimizing the error. This idea can be written as the following equation.

$$w_i = w_i - \alpha \frac{\partial C}{\partial w_i} \tag{26}$$

where in the above equation w_i is the ith weight, α is the learning rate, $\frac{\partial C}{\partial w_i}$ is the partial derivative of the cost function with respect to ith weight. The learning rate is used as a scale factor and it is an important hyperparameter. The gradient descent algorithm starts with initializing the weights randomly. After passing the forward path for the train data, the value of the loss is computed for each data and finally, the sum of the squared errors is calculated for each pair of input and output values. Then the partial derivative of the cost function with respect to each weight is computed and using the above equation, the weights are updated. This process is repeated until a minimum sum squared error is achieved or no further improvement is possible. It is easy to show that the partial derivative of the cost function with respect to each weight is equal to the following equation.

$$\frac{\partial C}{\partial w_j} = -\frac{2}{N} \sum_{i=1}^{m} x_j^{(i)} (y^{(i)} - \tilde{y}^{(i)}) \tag{27}$$

$$\frac{\partial C}{\partial b} = -\frac{1}{N} \sum_{i=1}^{m} (y^{(i)} - \tilde{y}^{(i)}) \tag{28}$$

where the x_i is the ith feature. From the above equation, you can get the gradient of the cost function in direction w_j. After computing the gradients of weights, you can update your linear regression model weights and bias term.

The linear regression with square error cost function is a convex optimization problem and does not have local minima. This fact can ensure for us that the gradient descent will reach the global minima if the learning rate is chosen wisely. The above formulation of linear regression has one hyperparameter which is α. The learning rate determines the size of the improvement step to take on each iteration of the procedure. If the value of the learning rate is too large, then the gradient descent takes a large step which causes to getting away from the local minima. In contrast, if the learning rate is too small, the gradient descent takes a small step. With this learning rate, the training procedure takes too long. Recall the insurance of reaching global minima in linear regression, we can guarantee that with a small learning rate, the linear regression reaches its global minima, but it takes too long. Therefore, it is important to select the learning rate carefully. There are some techniques to select the learning rate for each iteration so that it decreases after some iteration. These techniques can help to increase the speed of gradient descent and insurance to reaching to global minima.

The second approach to find the linear regression weights is to use the closed form equation. The linear regression problem has a closed form equation which results in the best weights. The closed form formula is given below.

$$w = (X^T X)^{-1} X^T y \qquad\qquad (29)$$

where the X is the matrix of training samples so that each row of X corresponds to a training sample, and each column corresponds to a single input dimension or simply a feature. The w weights are represented as a D-dimensional vector which is the dimension of a single input feature, and y is actual output values for all training samples. Notice that here y is vector so that each row is the actual prediction for each corresponding sample in X. In practice, it is useful to use gradient descent when you have a very large dataset either in the number of rows or the number of columns that may not fit into memory. For a large training dataset, the calculation of inverse matrix needs time and RAM. Therefore, it is not practical to use the closed form equation for these cases. Also, computing $(X^T X)^{-1}$ is roughly $O(n^3)$ where n is the number of input features. But in gradient descent, there is no need to compute the inverse matrix which causes fewer hardware resources. Also, if the training dataset is too large to load them in the RAM at once, there are other techniques to batch the train data and train linear regression. We will discuss this issue in the optimization section. The gradient descent is an iterative solution to solve the linear regression optimization problem while the closed form solution is non iterative. The other important difference between gradient descent solution and closed form solution is the learning rate hyperparameter. As we discussed before, you should choose the learning rate wisely, not too large that the gradient descent does not converge and not too small that the gradient descent needs very a long time to converge and reach its global minima.

Consider a polynomial linear regression model. One of the hyperparameters of this model is the degree of the polynomial. If this hyperparameter set is too low, the model may not learn the structure of the training dataset. Therefore, the model does not have good performance for new data. In other words, the model cannot be generalized. This is called the underfitting of the model. In contrast, if the degree of this model is set too large, the number of trainable weights increases. In this situation, the model may learn the structure of the data perfectly. Also the model train on the training dataset perfectly, it is not a good issue. Most of the time, the train and test data are noisy. Therefore, when a model trains on a noisy train dataset perfectly, it means that the model learns the structure of the noise. This phenomenon leads the model to have poor performance on the new data. This is called overfitting. We will discuss overfitting and underfitting with more details in the next chapter. When a model like polynomial linear regression is overfitting, it means that maybe the model has too much complexity. One approach to solve this problem is to reduce the complexity of the model which in our example means to decrease the polynomial degree. This approach is very time consuming and need resources for complex models and huge train dataset. Because you should decrease the degree of the polynomial, and train a polynomial linear regression with this degree, and validate it on the validation dataset. Therefore, it is not a common method to battle with model overfitting. The other approach is to use the regularization method. This technique adds a penalty term for each weight in the cost function so that the weights get close to zero. This leads to a weights vector with small values. In general, the regularization seeks to both minimize the sum of the squared error of the model on the training data but also to reduce the complexity of the model, like the number or absolute size of the sum of all coefficients in the model. Two popular examples of regularization procedures for linear regression are L1 regularization and L2 regularization. In L1 regularization the absolute sum of the coefficients is minimized while in L2 regularization the squared absolute sum of the coefficients is minimized.

The last issue that is good to know is the power of vectorization. Until now, for train a simple linear regression, the train dataset feed into the model sample by sample. This procedure is very time consuming and not logical. Because the For loops in any programming languages like Python slow down the running time, therefore we can vectorize the linear regression algorithm by expressing them in terms of vectors and matrices. This can speed up the training procedure significantly. Vectorization is the art of getting rid of explicit For loops in your code. Also, using vectorization technique has the following advantages. The first one is that the equations, and the code, will be simpler and more readable. When you program a machine learning algorithm like linear regression algorithm, the code is more clear and readable if use the vectorization concept instead of writing For loops. Most of the linear algebra libraries in any languages like Python are using a vectorization idea from concept to speed up their process. Also, matrix multiplication is very fast on a Graphics Processing Unit (GPU). Let's extend the simple linear regression problem to vectorized version. Consider you have m training samples which each of them is a n dimensional vector. The n also represents the number of input features or dimension of the feature space. We can put all the training sample in a matrix $m \times n$ which each sample stands in row. This can be written as the following

$$X = \begin{pmatrix} x^{(1)^T} \\ \vdots \\ x^{(m)^T} \end{pmatrix}_{m \times n} \tag{30}$$

In the above formula, the $x^{(1)}$ is representative of the sample number 1 in training example and is a row n dimensional vector. Remember that we use the uppercase format for matrix and lowercase format for the vectors. We can also put all the weights of linear regression in a column vector. If there are n input features, the dimension of the weights vector is $n \times 1$ and can be shown with the following formula

$$w = \begin{pmatrix} w_1 \\ \vdots \\ w_n \end{pmatrix}_{n \times 1} \tag{31}$$

where the w_i is the corresponding weight for the x_i input feature. This causes to the use of a simpler mathematic notation. Therefore, we can rewrite the linear regression forward pass as the following.

$$y = Xw + b\mathbf{1} \tag{32}$$

where $\mathbf{1}$ represents one vector which its elements are 1. If there are m training samples, we will expect the output variable or the predictions are m dimensional vector. If you familiar with linear algebra, it is easy to show that the above equation verifies our expectation. The above equation enables us to compute the output predictions in only one line of code. The bias term also can be put in the weights vector. This results in a simpler mathematic notation. To consider the bias term in weights vector, we should rewrite the input matrix as the following.

$$X = \begin{pmatrix} 1 & x^{(1)^T} \\ & \vdots \\ 1 & x^{(m)^T} \end{pmatrix}_{m \times (n+1)} \tag{33}$$

In the above input matrix, we add a one column vector. Now, we can rewrite the weights vector as the following.

$$w = \begin{pmatrix} w_0 \\ w_1 \\ \vdots \\ w_n \end{pmatrix}_{(n+1) \times 1} \tag{34}$$

In the above vector, the w_0 is the bias term. Again, we can rewrite the linear equation of linear regression as the following equation.

$$y = Xw \tag{35}$$

Note that the new definition of w and X are consistent with the original linear regression equation. Based on the vectorized equations of linear regression, we are able now to extend all other equations in linear regression algorithm like cost function in vectorized version. The vectorized equation of linear regression cost function can be written as the following.

$$C = \frac{1}{N}(y - Xw)^T (y - Xw) \tag{36}$$

where the $(y - Xw)^T$ is transpose version of $(y - Xw)$. If you are familiar with linear algebra, it is easy that this equation computes the loss for each sample, squares it and average on all of the loss. By using linear algebra, you can also extend the derivations to the vectorized version. The derivations of cost with respect to each weight and the bias term can be written as the following.

$$\frac{\partial C}{\partial w} = -\frac{2}{N} X^T (y - Xw) \tag{37}$$

$$\frac{\partial C}{\partial b} = -\frac{1}{N} (y - Xw) \tag{38}$$

First equation computes the derivation of each cost function with respect to each weight in only one line of code. After computing all derivations, the weights are updated using the following equation which is vectorized version of the original form.

$$w = w - \alpha \frac{\partial C}{\partial w} \tag{39}$$

Logistic regression

Logistic regression is one of the earliest machine learning algorithms. The name of this learning method can mislead some into thinking it is a regression algorithm while it is actually a classification algorithm. Logistic regression was developed and used by statisticians and after that, machine learning researchers used this function to build a logistic regression algorithm. This algorithm is mostly used in binary classification problems such as whether an email is a spam or not, whether the tumor is malignant or not, and whether there is a cat in the image or not. Albeit, this approach is simple and does not have enough power to handle the complex application in the real world, but it is the entrance of neural networks. By the end of this part, you will have a good intuition of logistic regression and will be ready to start neural networks.

The logistic regression is very similar to linear regression. The output variable in the linear regression model is a continuous variable that can vary from a very large positive number to a very negative number. This output is not suitable for classification problems for two reasons. The first regression is the nature of the output variable in the classification algorithms which is a discrete value while it continues in a linear regression. But someone may ask why doesn't it set a threshold on the output variable of linear regression to classify the input data. This output variable does not have meaning in the probability perspective. Therefore, it is not meant to set a threshold for the output variable of linear regression and classify the new data based on that. In logistic regression, this problem is solved by passing the output variable of linear regression from a function that can interpret its output as a probability. Therefore, the procedure of logistic regression is to use a linear combination of input features using some weights and pass the combined output of this linear equation. Finally, you decide that the new data belongs to the one class if the output is greater than 0.5 and if this value is less than 0.5, it belongs to the other class.

The core function in logistic regression is called regression function or sigmoid function. This function has some properties which can interpret its output as a probability value. For example, the output of the logistic function is always between 0 and 1. The logistic function formula is represented in the following.

$$h = g(z) = \frac{1}{1 + e^{-z}} \tag{40}$$

where e is the base of the natural logarithms and z is the actual numerical value that you want to map it into the range 0 and 1. In some literature, the output of the sigmoid function is also called activated. If you plot the above equation for the range -5 and 5 you will find out that logistic function has

S shape. Also, by increasing the z value, the output of logistic function will become close to 1. For example, if z is 5 then the output is 0.9933 which very close to 1. In contrast, by decreasing the z value, the output of logistic function will become close to 0. For example, if z is –5 then the output is 0.0067 which very close to 0. Remember that the logistic regression is a linear regression which uses logistic function to map input features to a probabilistic score. The main difference between linear regression and logistic regression is the output value. In linear regression it is numeric value but in logistic regression the output is binary.

The input values or input features are combined linearly using weights and add bias term to predict an output value. The linear equation in logistic regression is similar to the linear regression and can be written as the following:

$$z = wx + b \tag{41}$$

where w is the vector of weights and b is the bias term. Notice that x is just a single input. The z value can vary between infinity and negative infinity. By using logistic function, the z value maps to a probabilistic value (h). For example, let's say the h value for email spam problem is 0.9. This means there is 90% chance that the input email (x) is spam. Mathematically this can be written as

$$h = P(y = 1|x.w) \tag{42}$$

The following formula states the probability of $y = 1$ given x respect to w. Based on statistics, the probability of $y = 0$ given x respect to w can be concluded by the following equation:

$$P(y = 0|x.w) = 1 - P(y = 1|x.w) = 1 - h \tag{43}$$

To predict which class a data belongs to it, a threshold can be set. Based upon this threshold, the obtained estimated probability (the $P(y = 1|x.w)$ probability) is classified into one of two classes. In logistic regression the threshold is set to 0.5. Consider the spam classification. If the estimated probability of a new email is 0.67, then it classifies as spam.

Let's give an example to summarize concept of logistic regression. Consider a model that can predict whether a person is male or female based on their height. Given a height of 150 cm is the person male or female. As the input is a vector 1×1, the W is a vector 1×1 too. Assume W and b equals to 0.6 and –100, respectively. Using the above equations, we can calculate the probability of male given a height of 150 cm or more formally $P(y = male|x = 150.w = 0.6.b = -100)$.

$$z = wx + b = 0.6 * 150 - 100 = -10 \tag{44}$$

$$h = g(z = -10) = \frac{1}{1 + e^{10}} = 0.0000453978687 = 0 \tag{45}$$

The *h* value is less than 0.5 and therefore the logistic regression prediction is female. This example shows you how the logistic regression works.

Until now, you earn some intuition about logistic regression. But one may ask how the best values of *w* and *b* can be derived. The weights and bias term must be estimated from training data. There are two common ways to compute *w* and *b*. The first one is using closed form formulation. The main drawback in this approach is the curse of dimensionality because of inverse matrix computation in a closed form equation. The other approach is using the gradient descent optimization algorithm. In low dimension problems, the closed form equation has a lesser computation compared to gradient descent but in high dimensionality this statement does not hold true. Because of inverse matrix computation in first approach, the gradient descent needs less computation resource and time.

The gradient descent is an optimization method that find minima based on a measurement function. This measurement function in machine learning called loss function. Loss function is a function that tells us how good our model is. High loss for one sample can interpret that the model prediction does not match the right class and small loss means the model can classify the sample correctly. Remember that in machine learning problems there is some training data. Therefore, in each iteration of gradient descent, loss of each sample must be computed. To take account loss of all the training data, the mean of them must compute. This new function (average of losses) is called cost function. In gradient descent, the slope of cost will compute for each weight and then update each them in opposite direction slope.

$$w_i = w_i - \alpha \frac{\partial \cos t}{\partial w_i} \tag{46}$$

In linear regression, the loss function is mean squared error. By using this loss function, the optimization problem for logistic regression is non convex. Gradient descent will converge into global minimum only if the cost function is convex. If the cost function is non convex, it may stick in local minima. The loss function of logistic regression brings in the following equation

$$Loss = -y \times log(h(x)) - (1 - y) \times log(1 - h(x)) \tag{47}$$

In the above equation if $y = 1$, the second term is zero and the first term remain. If the prediction ($h(x)$) is 1, the loss term for this sample will be zero. But as the prediction is far from 1, the loss term will be larger and therefore can penalize model to train correctly. On the other hand, if $y = 0$, the first term will vanish and the second term will remain. In this case, as the prediction is far from 0, the loss term will increase and penalize the model. It is worth to notice that the negative function is because in training, we need to maximize the probability by minimizing loss function for all training samples. Decreasing the cost will increase the maximum likelihood assuming that samples are drawn from an identically independent distribution. It is easy to show that partial gradient for each weight is equal to:

$$\frac{\partial Loss}{\partial w_i} = (h - y)x_i \qquad (48)$$

where h is prediction probability, y is the actual output and x_i is the ith feature. The partial derivative for bias term is equal to:

$$\frac{\partial Loss}{\partial b} = (h - y) \qquad (49)$$

After computing the gradients, the weights and bias are updated. This implementation is for binary logistic regression. For data with more than 2 classes, softmax regression has to be used. In general, the softmax regression predicts the probability of being for each class. Therefore, the data belongs to a class with the most probability score. In neural networks section, the softmax regression will be explained.

Remember what we discuss about the vectorization concept in linear regression. This extension can use it the logistic regression algorithm. Using the vectorization idea, the training procedure is much faster than the original logistic regression training procedure. Also, this idea helps one to use resources efficiently.

K nearest neighbor classifier

The K Nearest Neighbor (KNN) is a supervised learning algorithm. This is one of the basic algorithms for classification. KNN is a non-parametric and lazy learning algorithm. Non-parametric means there is no assumption for underlying data distribution. In other words, the model structure is determined from the dataset. Also, this classifier does not need to train on a training dataset, and because of this, it is sometimes called a lazy classifier. More precisely, all training data is used in the test phase. This has some drawbacks and advantages which we can discuss. In the abstract, this classifier is memory-based and is not required to fit a model. In my opinion, this algorithm is inspired by the way of thinking of a child. Consider cat recognition. If you want to teach a child what is cat, you maybe will tell them the features of a cat can be their four legs. Afterward, when the baby sees an animal with four legs, recognizes it as a cat. This is due to the fact that the features of the cat are similar to the features of the animal (new observation). The K nearest neighbor classifier works in a similar procedure. In the abstract, the KNN algorithm assumes that similar things exist in close proximity. In other words, similar things are near to each other. The KNN classifier captures the idea of similarity with some mathematics like calculating the distance between points on a graph. In nutshell, the KNN classifier works like this: you have an existing set of example data with its class label, which is called the training dataset. When you give a new sample without a label, you compare that new sample to the all samples in the training dataset. Then, you take the most similar samples in the training dataset (the nearest neighbors) and look at their labels. Afterward, you look at the top K most similar samples from the training dataset. The value of the K is an integer and usually less than 20. Finally, you take a majority vote from the K most similar samples, and the majority is the new class you assign to the new sample you were asked to classify.

The similarity measurement is sometimes called distance, proximity, or closeness. Remember, there are many other ways of distance calculation. In general, the best distance calculation criteria are dependent on the problem. For example, in the face recognition area on the best similarity metric is cosine similarity. However, one of the most popular and familiar choices for distance calculation is the straight-line distance which is also called the Euclidean distance. The other important hyperparameter in the KNN classifier is K which indicates the number of nearest samples or neighbors. If the number of nearest samples (K) is too small for example 1, the output prediction will be confident if only the features are perfect. In contrast, if the K is too large, for example, 100, then samples from other classes stand in the proximity of this class. In nutshell, the selection of the K is very important in KNN. The KNN classifier has 7 steps which are listed below.

1. Normalize the features to have 0 mean and variance 1.
2. Initialize K to your chosen numbers of neighbors.
3. For new observations:
   3.1 Calculate the distance between the train examples and the current observation.
   3.2 Add the distance and the index of the example to an ordered collection.
4. Sort the ordered collection of distances and indices from smallest to largest (in ascending order) by the distances.
5. Pick the first K entries from the sorted collection.
6. Get the labels of the selected K entries.
7. If classification, return the mode of the K labels.

In the first step, you should prepare your trained dataset to fit the KNN model. For this classifier, the fit or training phase is equivalent to the sort training dataset. The usual normalization method is to subtract the mean of features from each of them. This subtraction forces the training dataset to be zero mean. Afterward, the zero mean features are divided by the variance of the train features so that the training dataset has unit variance. This normalization is due to the fact that each feature can vary differently and therefore cause the classifier to train slower. This issue can also mislead the classifier. In the second step, you should initialize the value of K or nearest neighbors. One approach for finding the best value of K is to train the KNN on the training dataset and validate it on the validation dataset and plot the accuracy or error of validation for each K. The general procedure of this plot for classification error is descending until a specific point and after that, the error ascending. This point is the best value for K. In the third step, the distance between the new observation and training dataset is computed. All of these distances and corresponding labels will be stored in a list. In the next step, these distances are sorted in a descending way. Now, the first K elements of the list are the samples with the lowest distance. In other words, they are nearest neighbors of this new observation. Finally, using the majority voting will determine the class of this sample.

As we discussed before, one of the important hyperparameters for KNN classifier is the number of neighbors or K. To select the best K for your dataset, you can run the KNN classifier several times with different values of K and choose the K that has the highest prediction accuracy on the test or validation dataset. We will discuss in detail the training, validation, and test dataset. It is common to vary the K value from 1 to 20 and find the best one. The distance criteria is also another hyperparameter. To select the best distance metric, you should know your problem, and based on that, choose one distance metric.

There are some tips that can give you more intuition about the KNN classifier. As the value of K decreases to 1, the classifier's predictions become less stable. Assume K is equal to 1 and your test or validation dataset is noisy. If the features are perfect, samples of each class stay in proximity of each other, but it is possible for a few noisy samples stand in this proximity. Therefore, these noisy samples can mislead the classifier when you run the KNN classifier for K equals 1. This is due to the fact that it is possible that the nearest neighbor of this sample is one of those noisy samples and results in the wrong prediction. In contrast, as the value of K increases, the KNN classifier predictions become more stable due to majority voting, and thus, more likely to make more accurate predictions until to a certain value. Eventually, in a certain value of K the accuracy of KNN classifier decreases or equivalently, the error rate increases. It is this point that can be considered as the best value of K. Note, it is best that K has the odd number. This is due to taking a majority vote among labels. The second issue is that KNN performs better with a low dimension of features vector than a large dimension of features vector. An increase in dimension also leads to the problem of overfitting. To avoid overfitting, the needed data will need to grow exponentially as you increase the number of dimensions. This problem of the higher dimension is known as the Curse of Dimensionality. We will discuss overfitting in the next chapter. To deal with the problem of the curse of dimensionality, you can use principal component analysis (PCA) before applying any machine learning algorithm, or you can also use the feature selection approach. It is shown that in large dimensions, the Euclidean distance is not useful anymore. Therefore, you can prefer other measures such as cosine similarity, which gets decidedly less affected by high dimensions. For example, in the modern deep learning method for face recognition, cosine similarity is a common distance calculation.

The KNN classifier is simple and easy to implement. Also, there is only one hyperparameter which makes the tuning simple. In contrast to neural networks, the KNN classifier has the same structure and does not need to build a model. The other advantage of this classifier is that it can be used for classification, regression, and searching. The independence of the training dataset is one of the important advantages of the KNN classifier. Despite the advantages of KNN, this classifier gets significantly slower as the number of examples increases. This is due to the training procedure of KNN. In KNN, the training dataset is the only sort, and as a result, it is very fast in training. But to test a new observation, the distance between all training observations and this one should be computed. This makes the KNN slow. In the abstract, the KNN classifier is fast in training but in the testing phase, it is slow and has many costs. Cost in the testing phase means time and memory. KNN's main disadvantage of becoming significantly slower as the number of data increases makes it an impractical choice in environments where predictions need to be made rapidly. Moreover, there are faster algorithms that can produce more accurate classification and regression results. However, if you have sufficient computing resources to speedily handle the data you are using to make predictions, the KNN classifier can still be useful in solving problems that have solutions that depend on identifying similar objects. An example of this is using the KNN classifier in recommender systems, an application of KNN-search.

Naïve Bayesian classifier

Naïve Bayesian classifier (NB classifier) is one of the most popular supervised machine learning algorithms that was inspired by the Bayes theorem of conditional probability to predict the class of a new observation. This is a probabilistic classifier that makes classifications using the Maximum a Posteriori decision rule in a Bayesian setting. It can also be represented using a very simple Bayesian network. The Naïve Bayesian classifier uses Bayes theorem of conditional probability assuming the predictor variables are independent. It also assumes the continuous variables to follow a normal distribution. In this algorithm, the classifier needs some assumptions on data while some other classifiers like the K nearest neighbor does not need any assumption about the training dataset. Naïve Bayes classifier is a simple algorithm and works very well if its assumption is observed, which is that the predictor variables are independent to each other. In the absence of large data, the algorithm sometimes work more efficiently rather than other complex classification algorithms like neural networks. This technique is more useful when it is more important to know the most likely class rather than knowing the actual probabilities for various classes. One example of successful applications of this algorithm is in the applications of classifying a mail as spam or not.

The goal of any probabilistic classifier with a feature vector is to determine the probability of the feature vector occurring in each class, and return the most likely class. Therefore, the conditional probability of each class given this feature vector should be computed. Before we explain the Naïve Bayesian theorem, let's discuss the Bayes theorem. Using the Bayes theorem, you can find the probability of A happening, given that B has occurred. You can write the above statements in the mathematics as the following:

$$P(A \mid B) = \frac{P(B \mid A)P(A)}{P(B)} \qquad (50)$$

In the above equation, B is the evidence and A is the hypothesis. The assumption made here is that the predictors or features are independent. That means the presence of one particular feature does not affect the other. Therefore, it is called naïve. Let's consider playing golf for example. Assume there are four weather conditions, humidity, temperature, and windy features. The main goal of this classifier is to predict whether it is a suitable day to play golf or not given the climates features. If the weather condition is rainy, the temperature is hot, the humidity is high and the day is not windy, it is not a suitable day for playing golf. There are two assumptions here, one as stated above these predictors or features are considered independent. In our example it means that if the temperature is hot, it does not necessarily mean that the humidity is high. Another assumption made here is that all the predictors have an equal effect on the outcome. That is, the day being windy does not have more importance in deciding to play golf or not. According to this example, Bayes theorem can be rewritten as the following equation:

$$P(y \mid X) = \frac{P(X \mid y)P(y)}{P(X)} \qquad (51)$$

In the above equation, the variable y is the class variable or label, which in our example represents whether it is a suitable day to play golf or not given the conditions. Variable X is the features. In the above example, the X

features consist of weather conditions, humidity, temperature, and wind. The *X* features can be written as the following vector:

$$X = (x_1.x_2. \ldots .x_n) \tag{52}$$

The $x_1, x_2, \ldots x_n$ represent the features, i.e., they can be mapped to outlook, temperature, humidity and windy. The n subscript denotes the number of features. For example, in the above example, the number of features is 4 therefore n equal to 4. By substituting for X and expanding, using the chain rule the original Bayesian equation converts to the following equation:

$$P(y \mid x_1.x_2 \ldots .x_n) = \frac{P(x_1 \mid y)P(x_2 \mid y) \ldots P(x_n \mid y)P(y)}{P(x_1)P(x_2) \ldots P(x_n)} \tag{53}$$

In the above equation, the first assumption of Naïve Bayesian for which the independence of features is used. For those whom that does not familiarise with independency rule in probabilistic, this assumption breaks the joint probability to the product of each events. Now, you can obtain the values for each by looking at the dataset and substituting them into the equation. Note that for all entries in the dataset, the denominator does not change and it remains static. Therefore, the denominator can be removed and a proportionality can be introduced as the following equation:

$$P(y \mid x_1.x_2 \ldots .x_n) \propto P(x_1 \mid y)P(x_2 \mid y) \ldots P(x_n \mid y)P(y) = P(y) \prod_{i=1}^{n} P(x_i \mid y) \tag{54}$$

In our case, the class variable *y* has only two outcomes, play golf or not. There could be cases where the classification could be multivariate. Therefore, you need to find the class *y* with maximum probability. For this, you can use the following equation.

$$y = argmax_y P(y) \prod_{i=1}^{n} P(x_i \mid y) \tag{55}$$

This equation is referred to as the Maximum a Posteriori decision rule. This is because, referring back to our formulation of Bayes rule, only the $P(B|A)$ and $P(A)$ terms are used, which are the likelihood and prior terms, respectively. If only the $P(B|A)$ term is used, the likelihood, would be using a Maximum Likelihood decision rule. By using the above equation, you can obtain the class, given the predictors. You can easily understand there is very little explicit training in Naive Bayes compared to other common classification methods like neural networks, logistic regression, and support vector machine. The only work that must be done before prediction is finding the parameters for the individual probability distributions of features, which can typically be done quickly and deterministically. This means that Naïve Bayesian classifiers can perform well even with high-dimensional data points or a large number of data points. Therefore, before training a Naïve Bayesian classifier, you should determine the conditional probability distribution. Each of these distributions have some parameters which should estimate from the training dataset. There are different types of Naïve Bayesian classifiers of which we point to three of them.

1. Multinomial Naïve Bayesian:

This is mostly used for document classification problems. For example, it classifies whether a document belongs to the category of sports, politics, technology, etc. In this example, the features or predictors used by the Naïve Bayesian classifier are the frequency of the words present in the document.

2. Bernoulli Naïve Bayesian:

This is similar to the multinomial Naïve Bayesian but the predictors are boolean variables. The parameters that we use to predict the class variable take up only values yes or no, for example if a word occurs in the text or not.

3. Gaussian Naïve Bayesian:

When the predictors are continuous values and are not discrete, you can assume that these values are sampled from a Gaussian distribution. Therefore, the formula for conditional probability distribution of each of the features can be written as the following:

$$P(x_i \mid y) = \frac{1}{\sqrt{2\pi\sigma_y^2}} \exp\left(-\frac{x_i - \mu_y}{2\sigma_y^2} \right) \tag{56}$$

Until now, you were familiar with the concept of Naïve Bayesian. One may ask how to train for example a Gaussian Naïve Bayesian model. What is the training procedure of this model? The training procedure of Naïve Bayesian model is simple and very fast. Consider the Gaussian Naïve Bayesian model. This model has 2 unknown parameters for conditional probability of each class which should estimate from the train dataset. In general, the Gaussian Naïve Bayesian model has 2 m trainable parameters, where m is the number of classes. Therefore, the training phase in this model is to estimate the μ_y and σ_y parameters for each class. The μ_{yi} and σ_{yi}^2 is the mean and variance of the ith class. To estimate these values, you need to estimate the mean and variance of each class from their trained data. Consider the ith class has k train samples. Therefore, you can estimate these values using the following equations:

$$\mu_j = \frac{1}{k}\sum_{l=1}^{k} x_l \tag{57}$$

$$\sigma_j^2 = \frac{1}{k}\sum_{l=1}^{k} (x_l - \mu_j)^2 \tag{58}$$

When you estimate your Gaussian Naïve Bayesian classifier, it is ready to use it in your application. Note that the mean and variance of each class are estimated only once from the training dataset. In the test phase, you should use the estimated values from the training phase.

Naïve Bayesian algorithms are mostly used in sentiment analysis, text classification, spam filtering, and recommendation systems. They are fast and easy to implement but their biggest disadvantage is the requirement of predictors or features to be independent. In most of the real life cases, the predictors are dependent, this hinders the performance of the classifier. Remember that the Naïve Bayesian classifier uses the pattern and structure of input data. Therefore, it is dependent to the training dataset [12, 20].

Artificial neural networks

An Artificial Neural Networks (ANN) loosely models the neurons in a human brain. This model is based on a collection of connections between units or nodes called "artificial neurons." Each connection from one artificial neuron to another, like the synapses in a biological brain, can transmit information or a "signal." Each of these connections is called "edge". An artificial neuron can process the received signal and then signal other artificial neurons connected to it. In common implementations of ANN, there is a real number as the signal at a connection between artificial neurons. The output of each one is calculated by some non-linear functions of the sum of its inputs. Any neurons and edges have a weight that adjusts as learning proceeds. The strength of a signal at each connection can increase or decrease due to the weight. Artificial neurons could have a threshold. Typically, artificial neurons are aggregated into layers. The transformation of the inputs in each layer may be different from others. The first layer is the input layer, and the signals feed on the first layer to the last layer (the output layer). It is possible for the signal to traverse the layers multiple times.

In the beginning, the ANN aimed to solve the same as the human brain. However, over time, the approach focuses on performing specific tasks, which cause deviations from biology. Artificial neural networks have been used on many tasks, including computer vision [41], speech recognition [42], machine translation [43], but it is not limited to these tasks.

In Deep Learning, we have multiple hidden layers in an artificial neural network. The effort of this approach is to model the human brain's process of light to vision and sound to hear. There are successful applications of deep learning like computer vision and speech recognition [44]. Deep learning is widely used in practical applications. In the next section, we mainly explain this model of machine learning.

The artificial neural network learning algorithm, or neural network, is a computational learning system that uses a network of functions to understand and translate a data input of one form into a desired output, usually in another form. Neural networks are one of the greatest learning algorithms that has significant effect on the artificial intelligence, and special machine learning areas. The great performance of neural networks, convince researchers and developers to use these networks in many applications and fields. Neural networks are being applied to many real-life problems today, including speech and image recognition, spam email filtering, finance, and medical diagnosis, to name a few. The neural networks can be put in supervised and unsupervised categories. In the supervised learning algorithms, they can be put in both regression and classification groups. One of the great features in neural networks is that the neural networks generally do not need to consider some specific rules that define what to expect from the input. The neural networks learning algorithm instead trains itself from processing many examples to learn what characteristics of the input are needed to construct the correct output. The data can have labels (in supervised learning algorithms) or do not have labels (in unsupervised learning algorithms). Once a sufficient number of examples have been processed, the neural network can begin to process new, unseen inputs and successfully return accurate results. The more examples and variety of inputs the program sees, the more accurate the results typically become because the program learns with experience.

Let's take an example to give you better intuition about neural networks. Imagine the simple problem of trying to determine whether or not an image contains a cat. While this is rather easy for a human to find out, it is much more difficult to train a computer to identify a cat in an image using classical methods. This is because the image is consisting from many numbers that put together in a matrix. Considering the diverse possibilities of how a cat may look in a picture, writing code to account for every scenario is almost impossible. Therefore, the machine learning algorithms should discover the structures in the data so that they can predict the desired output. In this example, the neural network should discover that a cat has four legs and many other features like this. In abstract, the neural network in the first layers learn the low level features and as the data pass them to the deeper layers, these features are combined and make new high level features that can classify what is the cat. Using several layers of functions to decompose the image into data points and information that a computer can use, the neural network can start to identify trends that exist across the many, many examples that it processes and classify images by their similarities.

The area of Neural Networks has originally been primarily inspired by the goal of modeling biological neural systems, but has since diverged and become a matter of engineering and achieving good results in machine learning tasks like classification and detection problems. In the biological neural system, many neurons are connected together and process the input data. For example, the visual neurons give the luminance of the environment and process this information. The output result of these process is that we can easily classify all objects around and so on. Therefore, it is suitable to start the neural networks section with a very brief and high level description of the biological neural system.

The basic computational unit of the brain is a neuron. Approximately 86 billion neurons can be found in the human nervous system and they are connected with approximately $10^{14} - 10^{15}$ synapses. Each neuron receives input signals from its dendrites and produces output signals along its single axon. Also, a neuron can have many dendrites but only one axon. In other words, a neuron can have many inputs that each of them is a feature and one output which is the processed result of the neuron. The axon eventually branches out and connects via synapses to dendrites of other neurons. Therefore, they form a network. In the computational model of a neuron, the signals that travel along the axons which can interpret as the feature input for the next neuron (called x_0) interact multiplicatively with the dendrites of the other neuron based on the synaptic strength at that synapse. The synaptic strength in the mathematic model means the weight of the neuron, which is called w_0. Therefore, the mathematic model for this input feature (x_0) and its corresponding weight (w_0) is equal to $x_0 w_0$. The synaptic strengths (the weights w) can control the contributions of each input features. As a result, the synaptic strengths (the weights w) are learnable and can control the strength of influence of one neuron on another. In the basic model, the dendrites carry the signal to the cell body where they all get summed. If the final sum is above a certain threshold, the neuron can fire or become active, sending a spike along its axon. In the computational model, we can assume that the precise timings of the spikes do not matter, and that only the frequency of the firing communicates information. Based on this rate code interpretation, we model the firing rate of the neuron with an activation function f, which represents the frequency of the spikes along the axon. Historically, a common choice of activation function is the sigmoid function σ, since it takes a real valued input and squashes it to a range between 0 and 1. In other words, one neuron has many features. These features are combined linearly in the neuron and pass through the activation function. Therefore, a neuron can model with a logistic regression classifier. Recall the logistic regression algorithm, there are some input features which are combined linearly. For a decision on whether this data belongs to which class, the combined features pass through a sigmoid function to squash between 0 and 1. A biological neural network is the combination of many neurons in a graph. Neural networks can model as collections of neurons that are connected in an acyclic graph. In other words, the outputs of some neurons can become inputs to other neurons. Cycles are not allowed since that would imply an infinite loop in the forward pass of a network. Neural network models are often organized into distinct layers of neurons. For regular neural networks, the most common layer type is the fully connected layer in which neurons between two adjacent layers are fully pairwise connected, but neurons within a single layer share no connections. Therefore, we can create an artificial neural network models by combining the logistic regression models and put them together in distinct layers.

To specify the neural network architecture, we should specify the number of layers and the number of neurons in each layer. Note that in some literature, the neural network is called by the number of its layers and do not consider the input layer. For example, when we have a neural network with 3 layers and an input layer, we can say this network is a 3-layer neural network. Therefore, a single layer neural network describes a network with no hidden layers and the input is directly mapped to the output. In this case, a single layer neural network is a special case of neural network which is a logistic regression model. A neural network consists of three parts. The first part is the input layer which include the input features that you want to feed to your neural network. The number of neurons in this layer is equal to the number of features. Remember that in some literature, the input layer is not considered as a layer. The second part is the hidden layers. Hidden layers are the layers between the input layer and output layer. The number of hidden layers is one of the neural network hyperparameters. The third layer is the output layer which give us the prediction of our model. The number of neurons in the output layer is equal to the number of classes. For example, if in object recognition problems with 10 objects, the output layer has 10 neurons which specify the probability of being each class. Also unlike all the layers in a neural network, the output layer neurons most commonly do not have an activation function. This is because the output layer is usually representing the class scores. Therefore, in most neural networks, the output layer uses the softmax function, which is sometimes called softmax layer.

The number of neurons and the number of layers are the two metrics that people use to measure the size of a neural network. These metrics identify the number of parameters for a neural network. Consider the following image. The network shown in the image is a 3-layers neural network. This neural network has one input layer with 3 features or inputs, two hidden layers where each of them has 4 neurons, and finally one output layer with one output neuron. Based on the network architecture, the output of this network has 2 classes. For example, is the email is spam or not, is this a cat image or dog image. All the layers in this network are a fully connected layer, because each neuron in each layer is connected to all neurons in the next layer. This network has 3 layers and the number of neurons from input layer to the output layer is equal to three, four, four, and one, respectively. Now let's compute the number of trainable parameters for this network. The input layer has three input features which they are fully connected to the first hidden layer with 4 neurons. Therefore, the first hidden layer has 12 weights. Also, each neuron in the first hidden layer has the bias term. As a result, the first hidden layer has 16 trainable parameters which include the weights and bias terms. The second hidden layer has 4 inputs from the previous hidden layer (the first hidden layer) and also it has 4 neurons. Therefore, the second hidden layer has 20 trainable parameters. The third layer is output layer which has 4 input from the second hidden layer. This layer also has only 4 trainable parameters which only include weights and not bias term. Remember that it is not usual to consider a bias term for the output layer. The total number of trainable parameters for this network equal to 40. The number of trainable parameters for each network can also represent the complexity of the neural network. To give you some intuition, modern Convolutional Networks contain orders of 100 million parameters and are usually made up of approximately 10 to 20 layers. When a neural network is called deep neural network that means it has lots of hidden layers. Usually, if the number of hidden layers is in excess of three, the network is called a deep neural network.

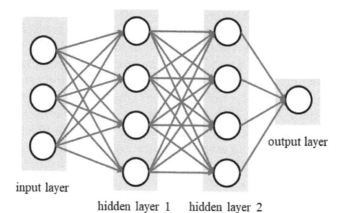

input layer

hidden layer 1   hidden layer 2

output layer

Figure 3.1. Example of three layer neural network. This network has two hidden layers, one input layer and one output layer.

Let's take a look in more details for the first hidden layer of the above network. Each of the neurons in the first hidden layer acts like a logistic regression classifier. Therefore, we can use the vectorized version of logistic regression equation for one neuron of this layer.

$$a_j^{(1)} = f(xw_j^{(1)}) \tag{59}$$

In the above equation, x is an input sample, $w_j^{(1)}$ is the weights which includes the bias term for the jth neuron in the layer (1), and the $a_j^{(1)}$ is the output of the jth neuron in the layer (1) after passing through the activation function f. This is a formula for one neuron in the first layer. In general, we can write the equation of neuron j in the layer i as the following

$$a_j^{(i)} = f(xw_j^{(i)}) \tag{60}$$

Before talk about other issues, let's extend the above equation with the vectorization concept. This extension enables you to compute the output of all neurons for layer i instantly. First, consider the first layer. In this layer, the input features are a matrix with $m \times n$ dimension which m is the number of train samples and n is the number of input features which in our example is 3. Remember that each w_j is a column vector with dimension $n \times 1$. We can put all the w_j of each neuron in a matrix in which each column represent the weights of a neuron like the following formula for our example.

$$W^{(1)} = (w_1^{(1)} \; w_2^{(1)} \; w_3^{(1)} \; w_4^{(1)}) \tag{61}$$

In the above equation, each $w_j^{(1)}$ is a column vector which includes the weights parameters and bias term. Note the dimension of $W^{(1)}$ which is a matrix with $n \times 4$ where the n is input features and the 4 is the number of neurons in the first layer. In general, the $W^{(i)}$ weights matrix for the *ith* layer has row numbers equal to the number of input features or input neurons (the number of neurons in the previous layer) and has column numbers equal to the number of neurons in this layer. For example, in the above network the $W^{(1)}$ has dimension of 3×4 where 3 is the number of neurons in the previous layer (input layer) and the 4 is the number of first layer. Therefore, we can write the forward equation of the first layer as the following.

$$H^{(1)} = XW^{(1)} \tag{62}$$

where the $h^{(1)}$ is the output for each neuron in the first layer before passing through the activation function. It is obvious that the dimension of the $h^{(1)}$ equal to $m \times 4$. The output of the first layer after passing the $h^{(1)}$ through the activation function is equal to the following equation.

$$A^{(1)} = f(h^{(1)}) \tag{63}$$

where the f is an element wise function. To compute the output for the second layer, the same procedure should be done. The weights matrix for the second layer is $W^{(2)}$ and its dimension equal to 4×4 and the third layer has the weights matrix $W^{(3)}$ with 4×1 dimension. Always remember that the forward pass of a fully connected layer corresponds to one matrix multiplication followed by a bias offset and an activation function. It is easy now to compute the output predictions using the forward pass with the above procedure for all samples.

The above equations, help us to write a closed form equation for each layer and also, program more readable and faster.

The main question in designing a neural network architecture is how do we select the best hyperparameters for our case. In other words, how do we decide on what architecture to use when faced with a practical problem? There are some questions that you should ask yourself before designing a deep neural network. These questions may help you to design a better architecture. Should I use no hidden layers? One hidden layer? Two hidden layers? How large should each layer be? Remember that as you increase the size and number of layers in a deep neural network, the capacity of the network will increase. This is due to the fact you combine many nonlinear neurons with each other and you can estimate any function with these powerful deep neural networks. In other words, the space of representable functions grows since the neurons can collaborate to express many different functions. For example, suppose you had a binary classification problem in two dimensions like is there cat and dog classification problem. You can train three separate shallow neural networks. The first network has 3 neurons, the second one has 6 and the third neural network has 20 neurons. Based on what we said, you expect the larger neural networks can represent more complicated functions. Therefore, the decision boundary of the first neural network is simple, the second neural network a little complex and the third neural network is more complex than the two other networks. In other words, the neural network with more neurons can represent more complicated functions. This property of the complex neural network can be good or bad. This can be good since the neural network learns to classify more complicated data and this is bad because of the overfitting phenomena. Overfitting occurs when a model with high capacity fits the noise in the data instead of the underlying relationship. For example, the model with 20 hidden neurons fits all the training data but at the cost of the learning noise of training data. The model with 3 hidden neurons only has the representational power to classify the data with a simple decision boundary. This decision boundary can be very similar to a linear decision boundary, for example it uses two or three lines to separate the data on which is a cat or a dog. The second neural network which has 6 neurons, can use a more complex decision boundary to separate data. In practice, the first and second neural networks have better generalization on the test (or validation) dataset. Therefore, it seems that smaller neural networks can be more suitable if the data is not complex. If the third neural network is used for simple data, the overfitting can occur. However, this is not the best solution to avoid overfitting. There are many other preferred ways to prevent overfitting in neural networks that we will discuss in the next chapter. In practice, it is always better to use these methods to control overfitting instead of reducing the number of neurons. The subtle reason behind this is that smaller networks are harder to train with local methods such as gradient descent optimization algorithm. It is clear that these neural network's loss functions have relatively few local minima, but it turns out that many of these minima are easier to converge to, which they are not suitable and have high loss and poor accuracy on the test dataset. Conversely, the bigger neural networks contain significantly more local minima, but these minima turn out to be much better in terms of their actual loss. Since the loss function of neural networks are not convex, it is hard to study these properties mathematically, but some attempts to understand these objective functions have been made. In practice, what you find is that if you train a small network the final loss can display a good amount of variance. On the other hand, if you train a large network you will start to find many different solutions, but the variance in the final achieved loss will be much smaller. In other words, all solutions are about equally as good, and rely less on the luck of random initialization. Therefore, it is more preferred to use larger and deeper neural networks instead of shallow neural networks with one or two hidden layers. To avoid overfitting in these deep neural networks, you can use some techniques like adding the regularization term in the loss function. The regularization term forces the neural network to have smaller weights. In nutshell, you should not be using smaller networks because you are afraid of overfitting. Instead, you should use as big of a neural network as your computational budget allows, and use other regularization techniques to control overfitting.

The other important issue in deep neural networks is the activation functions. In some literatures, the activation function is called nonlinearity function. Every activation function like sigmoid takes a single number and performs a certain fixed mathematical operation on it. In context of deep learning, there are several activation functions

1. Sigmoid activation function

The sigmoid function has the mathematical form $\sigma(z) = \dfrac{1}{1+e^{-z}}$. The sigmoid activation function takes a real value number and squashes it into a range between 0 and 1. In particular, large negative numbers become close to 0 and large positive numbers become close to 1. The sigmoid function has seen frequent use historically since it has a nice interpretation as the firing rate of a neuron, from not firing at all which is zero to fully saturated firing at an assumed maximum frequency which is 1. Recently, the sigmoid function has fallen out of favor and it is rarely ever used. This is due to the two major drawbacks. The first reason is that the sigmoid function saturates and kill gradients. A very undesirable property of the sigmoid neuron is that when the neuron's activation saturates at either tail of 0 or 1, the gradient at these regions is almost zero. In the following we will discuss more about backpropagation. The backpropagation is an algorithm to train a neural network. The main idea in backpropagation is chain rule of derivation. If one derivation has a value very close to zero, it forces the latter gradients get very close to zero and therefore the gradient of weights will vanish. In other words, during backpropagation this gradient will be multiplied to the gradient of this gate's output for the whole objective. Therefore, if the local gradient is very small, it will effectively kill the gradient and almost no signal will flow through the neuron to its weights and recursively to its data. Additionally, one must pay extra caution when initializing the weights of sigmoid neurons to prevent saturation. For example, if the initial weights are too large then most neurons would become saturated and the network will barely learn. The second reason is that the sigmoid outputs are not zero centered. This is undesirable since neurons in the later layers of processing in a deep neural network would be receiving data that is not zero centered. This has implications on the dynamics during gradient descent, because if the data coming into a neuron is always positive, then the gradient on the weights w will during backpropagation become either all be positive, or all negative. This could introduce undesirable zig-zagging dynamics in the gradient updates for the weights. However, note that once these gradients are added up across a batch of data, the final update for the weights can have variable signs, somewhat mitigating this issue. Therefore, this is an inconvenience but it has less severe consequences compared to the saturated activation problem above. The second reason leads the researcher to start using another activation function which is very similar to sigmoid function but is zero centered. This new activation function called Tanh activation function.

2. Tanh activation function

The Tanh activation function is S shape function which is very similar to sigmoid activation function. It squashes a real value number to the range -1 to 1. Like the sigmoid neuron, its activations saturate, but unlike the sigmoid neuron its output is zero centered. Therefore, in practice the Tanh activation function is always preferred to the sigmoid activation function. Also, it is easy to show that the Tanh neuron is simply a scaled sigmoid neuron, in particular the following holds equation

$$Tanh(z) = 2\sigma(2z) - 1 \tag{64}$$

3. Rectified Linear Unit (ReLU) activation function

The Rectified Linear Unit (ReLU) has become very popular in the last few years. The equation of ReLU activation function is written as the following.

$$f(z) = max(0, z) \tag{65}$$

The above equation states that the activation is simply compared with zero, if the value is greater than zero the output of ReLU is equal to the values otherwise the output is zero. There are several advantages and disadvantages to using the ReLUs activation function. One of the advantages of the ReLU activation function is that this function speeds up the training process. In other words, the Krizhevsky was shown that ReLU activation function can greatly accelerate the convergence of stochastic gradient descent compared to the sigmoid and Tanh activation functions by approximately factor 6. It is argued that this is due to its linear and non-saturating form. Recall the disadvantages of the sigmoid and Tanh activation functions, in these functions the gradient may vanish due to the saturation in large positive and negative numbers. The second advantageous of ReLU activation function is that it is compared to sigmoid and Tanh activation function that involve expensive operations like exponentials and division computation, the ReLU can be implemented by only a simple element wise comparison the matrix with zero threshold.

Also the ReLU activation function is very popular in the deep learning community, the ReLU units can be fragile during training and can die. For example, a large gradient flowing through a ReLU activation function can cause the weights to update so that the neuron will never activate on any data point again. If this happens, the gradient flowing through the unit will forever be zero from that point on. That is, the ReLU neurons can irreversibly die during training. For example, you may find that as much as 40% of your network can be dead if the learning rate is set too high. In other words, the ReLU neurons never activate across the entire training dataset. With a proper setting of the learning rate this is less frequently an issue. This problem is similar to the gradient vanishing. In overall, the ReLU activation function is one of the best and popular activation functions which it is used in most of the modern deep learning architecture.

4. Leaky ReLU activation function

Leaky ReLU activation function is attempted to fix the dying ReLU problem. The output value of ReLU activation function for negative input is zero. This can lead to the dying ReLU problem. Therefore, in leaky ReLU instead of the output of the ReLU function being zero for negative input values, it has a small positive slope like 0.01. The equation of leaky ReLU activation function is written as the following.

$$f(z) = \begin{cases} z & z \geq 0 \\ \alpha z & z < 0 \end{cases} \tag{66}$$

where in the above equation α is a small constant. The slope in the negative region can also be made into a parameter of each neuron.

5. Maxout activation function

Other types of neurons have been proposed that do not have the functional linear form (which is $f(w^T x + b)$) where an activation function is applied on the dot product between the weights and the data. One relatively popular choice is the Maxout activation function which is introduced by Goodfellow. This activation function generalizes the ReLU and Leaky ReLU activation functions. The Maxout activation function computes the following function.

$$f(z) = max(w_1^T z + b_1 \, . \, w_2^T z + b_2) \tag{67}$$

Notice that both the ReLU and Leaky ReLU activation function are a special case of this form. For example, in the ReLU activation function the weights (w_2) and bias term (b_2) are zeros. Therefore, the Maxout activation function has all the benefits of a ReLU activation function like the linear regime of operation, and no saturation and does not have its drawbacks like dying ReLU. However, unlike the ReLU neurons it doubles the number of parameters for every single neuron, leading to a high total number of parameters. Therefore, the computational amount in Maxout activation function is twice that of the ReLU activation function. It is worth noting that in some literature a neuron with Maxout activation function is also called Maxout neuron.

One may ask why we use the nonlinearity function (activation function) in the context of neural network. The first reason is that the most problems have a nonlinearity nature and therefore they cannot be estimated by the linear functions. Based on experiences, the performance of neural networks using nonlinear activation functions result better. The second reason is that when you use all neurons with a linear activation function, you can comprise all of them as a single neuron or logistic regression. Hence, to use the power of the neural networks we should use the nonlinear activation functions.

Until now you are familiar with the concept of neural networks and how they work. The other important sections in neural networks is how they train. Consider a large deep neural network which has millions of weights and biases, what is the best algorithm to train this neural network? The backpropagation algorithm is one of the best algorithms to train a neural network. Backpropagation in a neural network is a short form for backward propagation of errors. It is a standard method of training artificial neural networks. This method helps calculate the gradient of a loss function with respect to all the weights in the network. The backpropagation is the essence of neural network training. It is the practice of fine-tuning the weights of a neural network based on the loss which is obtained in the previous epoch (or equivalently iteration). Proper tuning of the weights ensures lower error rates, making the model reliable by increasing its generalization. The Backpropagation algorithm in neural network computes the gradient of the loss function for a single weight by the chain rule in derivation. It efficiently computes one layer at a time, unlike a native direct computation. It computes the gradient, but it does not define how the gradient is used. It generalizes the computation in the delta rule. In other words, the backpropagation algorithm starts with computing the gradient of the loss function of a neural network with respect to the previous layer. After that, the gradients of the second layer from the last respect to the gradients in the first layer from the last is computed. This procedure for computing the gradients of weights are used until the first input layer. This is the reason to call this algorithm backpropagation. Finally, the weights are updated using the gradient descent algorithm. There are two types of backpropagation, static backpropagation and recurrent backpropagation. The static backpropagation produces a mapping of a static input for static output. It is useful to solve static classification issues like optical character recognition. But the recurrent backpropagation is fed forward until a fixed value is achieved. After that, the error is computed and propagated backward. The mathematic details of the backpropagation are out of scope of this book and we ignore it.

The other important note in training deep neural networks is how to initialize the weights for the starting points. Previously, we mentioned that the neural network models are fit using an optimization algorithm called gradient descent that incrementally changes the network weights to minimize a loss function. During training we will hope that the resulting weights have a high performance. In other words, the optimization problem falls inside the local minima. This optimization algorithm requires a starting point in the space of possible weight values from which to begin the optimization process. Weight initialization is a procedure to set the weights of a neural network to small random values that define the starting point for the optimization of the neural network model. Remember that training deep models is a sufficiently difficult task that most algorithms are strongly affected by the choice of initialization. The initial point can determine whether the algorithm converges at all or not. With some initial points the algorithm encounters numerical difficulties and fails to train. In other words, with some initial points the gradient descent stuck in bad local minima which the model performance on the test (or validation) dataset is poor. When you train a neural network with different initial weights at the starting point, it is very probable that the performance of each model is different from the others, and therefore the output for a sample differs from the others. It is obvious that this is due to the random initialization which because the gradient descent is stuck in different local minima for each starting points and final weights for each model differs from the other ones.

Note that adding randomness can help avoid the good solutions and help find the really good and great solutions in the search space. They allow the model to escape local minima or deceptive local optima where the learning algorithm might get such, and help find better solutions, even a global minimum. One type of randomness is random initial weights. The other type of randomness that can help to train a better model and avoid overfitting is random shuffle of samples each epoch or iteration. The random initial weights allow the model to try learning from a different starting point in the search space each algorithm runs and allows the learning algorithm to break the symmetry during training network. However, the random shuffle of examples during training ensures that each gradient estimate and weight update is slightly different. The reason to use the random shuffling data is that the network can learn the arrangement of train dataset and result in a perfect accuracy on the train dataset but loses the generalization ability. This is causes the overfitting problem. But when the train dataset shuffles, the neural network is forced the structure of the pattern in the train dataset.

One may want to initialize all the weights of a neural network to the same value for example the zero value. This idea is not working because of the symmetry of the weights and their gradients. The gradients for all weights in this case are equal to zeros. Therefore, the equations of the training algorithm would fail to make any changes to the network weights, and the model will be stuck. It is important to note that the bias weight in each neuron is set to zero by default, not a small random value. Specifically, nodes that are in a same hidden layer and connected to the same inputs must have different weights for the learning algorithm to update the weights. This is often referred to as the need to break symmetry during training. If two hidden neurons with the same activation function are connected to the same inputs, we will expect each of them has different training procedure and this is when it initializes with different parameters. If they have the same initial parameters, then a deterministic learning algorithm applied to a deterministic cost and model will constantly update both of these units in the same way. Traditionally, the weights of neural networks with random initialization have small values. Good initialization of the weights of neural networks can speed up the training process and must be set carefully. There are some initialization techniques for neural network which the most important of them are glorot_normal, glorot_ uniform, he_nomal, and he_uniform.

The weights initialization method depends on the activation function. It is best to use one specific weight initialization for neurons we use the Tanh and sigmoid activation function and another initialization method for the neurons that use the ReLU activation function. The current standard approach for initialization of the weights of neural network layers and nodes that use the sigmoid or Tanh activation function is called glorot or xavier initialization. There are two versions of this weight initialization method, which we will refer to as xavier and normalized xavier. The author of this weights initialization are Glorot and Bengio. They proposed to adopt a properly scaled uniform distribution for initialization. This is called Xavier initialization. The basic idea behind the Xavier initialization is that it assumes the activation functions are linear. This assumption is invalid for ReLU and Leaky ReLU. Also both approaches were derived assuming that the activation function is linear, nevertheless, they have become the standard for nonlinear activation functions like Sigmoid and Tanh, but not ReLU. The Xavier initialization method is calculated as a random number with a uniform probability distribution between the range $-\frac{1}{\sqrt{n}}$ and $+\frac{1}{\sqrt{n}}$, where n is the number of inputs to the neurons. In other words, the value of n equal to the neurons in previous layer. For example if the third layer in a neural network has 10 neurons then the value of n for the fourth layer is 10 if they use Tanh or sigmoid activation function. This can be written as the following equation.

$$weight = Uniform\left[-\frac{1}{\sqrt{n}} \cdot +\frac{1}{\sqrt{n}}\right] \tag{68}$$

The implementation of the above equation is very simple in any programming languages like Python. The other version of Xavier weight initialization is normalized Xavier weight initialization. As you can guess from the name of this method, the weights are normalized. The normalized Xavier initialization method like the Xavier method is calculated as a random number with a uniform probability distribution but in the range of $-\frac{\sqrt{6}}{\sqrt{n+m}}$ and $+\frac{\sqrt{6}}{\sqrt{n+m}}$, where n is the number of inputs to the neuron or more specifically the number of neurons in the previous layer and m is the number of outputs from the layer or more specifically number of neurons in the current layer. The normalized Xavier initialization equation can be written as the following.

$$weight = Uniform\left[-\frac{\sqrt{6}}{\sqrt{n+m}} \cdot +\frac{\sqrt{6}}{\sqrt{n+m}}\right] \tag{69}$$

The Xavier weight initialization methods were found to have problems when used to initialize networks that use the ReLU activation function. Therefore, a modified version of the Xavier weight initialization was developed specifically for neurons and layers that use ReLU activation function which popular in the hidden layers of most multilayer Perceptron and convolutional neural network models. The current standard approach for initialization of the weights of neural network layers and neurons that use the ReLU activation function is called "he" weights initialization and introduced in the 2015. The "he" initialization method is calculated as a random number with a Gaussian probability distribution with a mean of zero and a standard deviation of $\sqrt{\frac{2}{n}}$, where n is the number of inputs to the neuron or more specifically the number of neurons in the previous layer. The equation of the he weights initialization is written as the following.

$$weight = Gaussian\left(0, \sqrt{\frac{2}{n}}\right) \tag{70}$$

Until now you have learned how to use neural network to predict the class of a new sample and how to train a neural network using a backpropagation algorithm. When you want to implement the training procedure of a neural network by yourself. The most difficult part of training procedure is backpropagation. Implementing backpropagation from scratch is usually more prone to bugs. Therefore, it is necessary before running the neural network on training data to check if your implementation of backpropagation is correct. The basic idea is check the real gradient with the gradient computed from the backpropagation. The real gradient could be estimated numerically. This algorithm is called gradient checking. Before discussing the matter of gradient checking, let's review the backpropagation. In the backpropagation loop over the neurons in the reverse topological order starting at the final neuron or the output layer to compute the derivative of the cost function with respect to each weight and bias. In other words, the derivative of cost function with respect to all weights $\frac{\partial J}{\partial w_i}$ and biases $\frac{\partial J}{\partial b}$. As told before, the algorithm to test your implementation is by computing numerical gradients and comparing it with gradients from backpropagation. If you are familiar with calculus, there are two methods to compute the numerical gradients two sided form and right hand form. The two sided form equation is written in the following.

$$\frac{J(w_j^{(i)} + \epsilon) - J(w_j^{(i)} - \epsilon)}{2\epsilon} \tag{71}$$

where in the above equation J is the cost function, ϵ is a very small number, and $w_j^{(i)}$ is the weight of neuron j in the layer j. The right sided form equation is written in the following.

$$\frac{J(w_j^{(i)} + \epsilon) - J(w_j^{(i)})}{\epsilon} \tag{72}$$

The two sided form of approximating the derivative is closer than the right hand form. Let's take an example and show why the two sided is more accurate. Consider the following function $f(x) = x^3$ by taking its derivative at $x = 2$. The analytical derivative of this function at the desired point is equal to the 12.

$$\nabla_x f(x) = 3x^2 \Rightarrow \nabla_x f(2) = 12 \tag{73}$$

The two sided numerical derivative estimation with $\epsilon = 0.01$ is equal to

$$\frac{f(x + \epsilon) - f(x - \epsilon)}{2\epsilon} = \frac{(2 + 0.01)^3 - (2 - 0.01)^3}{0.02} = 12.001 \tag{74}$$

While the right sided numerical derivative estimation with $\epsilon = 0.01$ is equal to

$$\frac{f(x + \epsilon) - f(x)}{\epsilon} = \frac{(2 + 0.01)^3 - 2^3}{0.01} = 12.006 \tag{75}$$

As you can see above, the difference between the analytical and two sided numerical derivatives is 0.01 however, the difference between analytical derivative and right sided derivative is 0.06. Therefore, it is best to use a two sided epsilon method to compute the numerical gradients. The two sided numerical derivative is also called centered numerical derivative. Now let's answer the question what are the details of comparing the numerical gradient (like two sided gradient) f'_n and analytic gradient (backpropagation) f'_a? You might be tempted to keep track of the difference $|f'_a - f'_n|$ or its square and define the gradient check as failed if that difference is above a threshold. However, this is problematic. For example, consider the case where their difference is equal to 1e–4. This seems like a very appropriate difference if the two gradients are about 1.0, therefore the two gradients are matching. But if the gradients were both on order of 1e–5 or lower, then the difference value which is 1e–4 can be a huge difference and likely a failure. Hence, it is always more appropriate to consider the relative error. The equation that uses the in gradient checking is written in the following

$$\frac{|f'_a - f'_n|}{\max(|f'_a|, |f'_n|)} \tag{76}$$

which considers their ratio of the differences to the ratio of the absolute values of both gradients. Note that normally the relative error formula only includes one of the two terms (either one), but it is preferred to max both to make it symmetric and to prevent dividing by zero in the case where one of the two is zero which can often happen, especially with ReLU and leaky ReLU activation function. However, one must explicitly keep track of the case where both are zero and pass the gradient check in that edge case. In practice, the value of the above equation can be checked as the following:

- If the relative error is greater than the 1e–2, it usually means the gradient is probably wrong and therefore there are some bugs in implementation of backpropagation.
- If the relative error is between 1e–4 and 1e–2, you should make it feel uncomfortable. In this case you must check your implementation of backpropagation because it is likely to be a small bug in your implementation.

- If the relative error is less than 1e–4, it is usually okay for objectives with kinks. But if there are no kinks like use of Tanh and sigmoid nonlinearities and softmax, then the value 1e–4 is too high.

- If the relative error is less than the 1e–7, then your backpropagation implementation is okay and there are not any bugs in it.

Also keep in mind that the deeper the network, the higher the relative errors will be. So if you are gradient checking the input data for a 10-layer network, a relative error of 1e–2 might be okay because the errors build up on the way. Conversely, an error of 1e–2 for a single differentiable function likely indicates incorrect gradient.

There are some practical notes in the gradient checking that you should consider to get a correct result. Be careful that you do not let the regularization overwhelm the data. In deep neural networks it is common to use the regularization technique, therefore the loss function is a sum of the data loss and the regularization loss like L2 penalty on the weights. One danger to be aware of is that the regularization loss may overwhelm the data loss, in which case the gradients will be primarily coming from the regularization term which usually has a much simpler gradient expression. This can mask an incorrect implementation of the data loss gradient. Therefore, it is recommended to turn off regularization and check the data loss alone first, and then the regularization term second and independently. One way to perform the latter is to hack the code to remove the data loss contribution. Another way is to increase the regularization strength so as to ensure that its effect is non-negligible in the gradient check and that an incorrect implementation would be spotted. The other tip in gradient checking is that to remember to turn off dropout regularization and augmentations. When performing gradient checks, remember to turn off any non-deterministic effects in the network, such as dropout regularization, and random data augmentations. Otherwise these can clearly introduce huge errors when estimating the numerical gradient. The drawback of turning off these effects is that you would not be gradient checking them. Therefore, a better solution might be to force a particular random seed before evaluating both $f(x + h)$ and $f(x - h)$, and when evaluating the analytic gradient. The next tip is that we only check a few dimensions. In practice the gradients can have sizes of million parameters. In these cases, it is only practical to check some of the dimensions of the gradient and assume that the others are correct. One issue that you should be careful of is to make sure to gradient check a few dimensions for every separate parameter. In some applications, people combine the parameters into a single large parameter vector for convenience. In these cases, for example, the biases could only take up a tiny number of parameters from the whole vector, so it is important to not sample at random but to take this into account and check that all parameters receive the correct gradients.

Now you are able to test and train a neural network. Also, you are equipped to check whether your backpropagation implementation is correct or not. However, the neural networks are very probable to overfitting. There are some techniques like regularization that can avoid the model to overfit. However, there are some other methods that can help you to monitor the model in the training process and find the best weights. One of these methods is babysitting the learning process. You can plot the value of loss of accuracy for each epoch. These plots are the window into the training process and should be utilized to get intuitions about different hyperparameter settings and how they should be changed for more efficient learning. The horizontal axis of these plots are always in units of epochs, which measure how many times every example has been seen during training in expectation. One epoch in neural network concept means that every example has been seen once. It is preferable to track epochs rather than iterations since the number of iterations depends on the arbitrary setting of batch size. One of the quantities that is useful to track during training is the loss, as it is evaluated on the individual batches during the forward pass. The vertical axis in this plot is the value of loss function after updating the weights and bias term in each epoch. You can understand a lot of information by monitoring the loss value at each epoch. With high learning rates the loss value decreases very fast. Higher learning rates will decay the loss faster, but it is very probable that they get stuck at worse values of loss. In the contrast, with the low learning rate, the value of loss decrease slowly. Lower learning rates will decay the loss slower, and it takes more time to converge the model and the loss value be at its minimum value. The good learning rate is a value that does not get stuck in the bad local minima and also is fast enough to train a neural network. We will discuss in more detail about how to set the learning rate in the future. Also note the last tip about the loss plot. When you train a neural network, it is probable that the loss value for some epochs decrease and after that the loss increases. At this point, in the model overfit and you should stop the training. This point has the best values of weights and biases. The amount of wiggle in the loss is related to the batch size. When the batch size is equal to one, the wiggle will be relatively high. When the batch size is the full dataset, the wiggle will be minimal because every gradient update should be improving the loss function monotonically unless the learning rate is set too high. Some researchers prefer to plot their loss functions in the log domain. Since the learning progress generally takes an exponential form shape, the plot appears as a slightly more interpretable straight line, rather than a hockey stick. The other important quantity to monitor during the training procedure is the validation and training accuracy. The vertical axis of this plot is the accuracy of validation and train dataset. This plot can give you valuable insights into the amount of overfitting in your model. The gap between the training and validation accuracy indicates the amount of overfitting. If the validation accuracy curves and trains accuracy curves have larger difference, you can conclude that a strong overfitting occurs for the model. When you see this in practice you probably want to increase regularization like a stronger L2 weight penalty, and more dropout, or collect more data for training. The other possible case is when the validation accuracy tracks the training accuracy fairly well. This case indicates that your model capacity is not high enough. Therefore, you should make the model larger by increasing the number of parameters. The last item that you may monitor for it, is the ratio of the update magnitudes to the value magnitudes. You might want to evaluate and track this ratio for every set of parameters independently. A rough heuristic is that this ratio should be somewhere around 1e–3. If it is lower than this, then the learning rate might be too low. If it is higher then, the learning rate is likely too high. Instead of tracking the min or the max, some researchers prefer to compute and track the norm of the gradients and their updates instead. These metrics are usually correlated and often give approximately the same results.

An incorrect initialization can slow down or even completely stall the learning process. Luckily, this issue can be diagnosed relatively easily. One way to do so is to plot activation or gradient histograms for all layers of the network. Intuitively, it is not a good sign to see any strange distributions like with Tanh neurons and you would like to see a distribution of neuron activations between the full range of [−1.1], instead of seeing all neurons outputting zero, or all neurons being completely saturated at either −1 or 1. The last tip about babysitting the learning process is when one is working with image pixels it can be helpful and satisfying to plot the first layer features visually. Noisy features indicate could be a symptom for example not converged network, improperly set learning rate, very low weight regularization penalty. In contrast, the nice, smooth, clean and diverse features are a good indication that the training is proceeding well.

When you train a neural network using the gradient descent optimization algorithm, one of the key hyperparameters is the learning rate. The learning rate scales the magnitude of weight updates in order to minimize the loss function of a neural network. Therefore, this parameter controls the speed of training phase. For example, if the learning rate is set too low, the training progress will take it very slowly since at each iteration of training, the weights are updated in a very tiny step. In other words, smaller learning rate may allow the model to learn a more optimal or even globally optimal set of weights but may take significantly longer to train. In contrast, if the learning rate is set too high the neural network can diverge since at each iteration the weights are updated in a very big step and therefore, can get lost to find the best weights. In other words, a large learning rate allows the neural network to learn faster, at the cost of arriving on a suboptimal final weights. With a good choice of learning rate, the model will estimate the best weights of the neural network for your problem. Unfortunately, there is not any analytically calculation formula to find the optimal learning rate for a given model on a given dataset. Instead, an appropriate learning rate must be discovered via trial and error. For learning rates which are too low, the loss may decrease. This reduction amount is very shallow rate and therefore needs more training epochs. Conversely, increasing the learning rate will cause an increase in the loss as the parameter updates cause the loss to bounce around and even diverge from the minima. In this case the model requires fewer training epochs. The best learning rate is associated with the steepest drop in loss. This is key fact to select the best learning rate in training. One of the methods which needs time and resources is to babysit the model. For example, you start your training with a relatively high learning rate in the first epochs. The model usually is far from the best global or local minima and therefore to speed up the training process, it is logical to set the learning rate with a relatively high value. This can help you to save your time. Until a specific epoch, the loss value will decrease and after that the loss value will increase. At this epoch your model gets lost in the feature space. Therefore, you should decrease learning rate. This causes the model's focus only on that area of feature space. This process will repeat until the best accuracy on validation and train dataset result while the loss of validation and train dataset are very small. Also, you can get a good result using this method, but this is very time consuming and needs resources. The other method which is better than the first idea is to set a schedule to adjust your learning rate during training. This method recommends to start training with a relatively high learning rate and then gradually lowering the learning rate during training. The intuition behind this approach is that it is best to traverse quickly from the initial parameters to a range of good parameter values but then the learning rate reduces enough that the model can explore the deeper, but narrower parts of the loss function. This is also the intuition of the babysitting idea. The most popular form of learning rate scheduling is a step decay where the learning rate is reduced by some percentage after a set number of training epochs. Some of the most popular learning rate scheduling is listed in the following

• Step decay

One of the popular learning rate schedules used with deep learning models is to systematically drop the learning rate at specific times during training. Usually, the step decay method is implemented by dropping the learning rate by half every fixed number of epochs. For example, we may have an initial learning rate of 0.1 and drop it by 0.5 every 10 epochs. The first 10 epochs of training would use a value of 0.1, in the next 10 epochs a learning rate of

0.05 would be used, and so on. If you plot the learning rates for this example up to 100 epochs, you get a graph like stairs. In the first epochs, the distance between two learning rate are large but as the model trains in more epochs, the decreasing of learning is reducing.

• Time decay

Time based learning rate decay is the other important method for setting this hyperparameter during the training phase. This method decreases the learning rate in each epoch. This method has the following mathematical form

$$\alpha = \frac{\alpha_0}{1 + decay \times epoch} \tag{77}$$

where in the above equation the α_0 is initial learning rate, *decay* is learning rate decay in each epoch, and *epoch* is represent the time. In this equation, if the *decay* is equal to 0, then the α equals to α_0 in every epochs. When the *decay* has a high value, then the learning rate will decrease fast. Also, if the *decay* has a low value, then the learning rate will decrease slowly. The best approach is to set a suitable *decay* and α_0 so that the learning rate decrease smoothly.

• Exponential decay

The other important learning rate scheduling method is exponential decay. In this method, the learning rate decrease exponential respect to time. This method has the following mathematical form

$$\alpha = \alpha_0 \, e^{-decay \times epoch} \tag{78}$$

where in the above equation the α_0 is the initial learning rate, *decay* is learning rate decay in each epoch, and *epoch* is represent the time. In this equation. Note that when the *decay* parameter equals to 0, then the α equals to α_0 in every epochs [1, 12, 20].

Decision trees and random forests

Decision trees are used for classification and also regression as a model. We can use trees to answer sequential questions, which send us down a certain route of the tree given the answer. "If this, then that" condition is the behavior of the model that, in the end, gives the result. Building a decision tree requires algorithms capable of determining an optimal choice at each node. The Hunt's algorithm [45] is the well-known one. This algorithm makes the optimal decision at each step, not considering the total optimum. It means that at each step, the algorithm selected the best answer. Though, choosing the best result at a given step does not guarantee that you reach optimal decision down from route to the final node of the tree, leaf node [46].

Decision trees, especially the deep ones, are prone to overfitting because the amount of the desired specificity may lead to a smaller sample of events that meet the previous assumptions. This small sample could generate unreliable results. Ideally, we would like to minimize errors on account of both bias and variance. In random forests, we rarely encounter overfitting. A random forest is simply a collection of decision trees. The results of trees are aggregated to reach the final result. The ability to limit the overfitting in random forests without substantially increasing bias error made them powerful models [12].

Support vector machines

Support vector machines (SVMs) [47], or support vector networks, are used for classification and regression. SNMs are a set of related supervised learning methods. Considering a set of training data, each labeled data that represents the membership of two categories; a model is built by an SVM training algorithm. This model can predict that a new example belongs to which category. An SVM training algorithm is a non-probabilistic, binary, linear classifier, but there are some methods such as Platt scaling to set and apply the SVM for probabilistic classification. Additionally, to implement linear classification in SVMs, it can use the kernel trick, implicitly mapping their inputs into high-dimensional feature spaces.

Until now, you will accustom with linear regression and logistic regression [2] algorithms. In linear regression the output variable continues while in the logistic regression the output variable is discrete. In my opinion, understanding some simple classification and regression algorithm can help you to understand the support vector machine classifier and support vector regression. Support vector machine is another simple algorithm that every machine learning expert should know. Support vector machines are highly preferred by many as it produces significant accuracy with less computation power. Support Vector Machine, abbreviated as SVM can be used for both regression and classification tasks. But, it is widely used in classification objectives. The regression version of SVM is called Support Vector Regression (SVR).

In general, the objective of the support vector machine algorithm is to find a hyperplane in an N dimensional space that distinctly classifies the data points. The N comes from the feature vector dimension. To separate the two classes of data points, there are many possible hyperplanes that could be chosen. The objective of the SVM is to find a plane that has the maximum margin. In other words, the goal of the SVM is to find a plane that has the maximum distance between the data points of both classes. Maximizing the margin distance provides some reinforcement so that future data points can be classified with more confidence. Before we dive into more detail, let's introduce hyperplane and support vectors. Hyperplanes are decision boundaries that help to classify the data points. Data points falling on either side of the hyperplane can be attributed to different classes. Also, the dimension of the hyperplane depends upon the number of features. If the number of input features is equal to two, then the hyperplane is just a line. If the number of input features is equal to three, then the hyperplane becomes a two dimensional plane. It becomes difficult to imagine when the number of features exceeds three. Support vectors are data points that are closer to the hyperplane and influence the position and orientation of the hyperplane. Using these support vectors, the margin of the classifier is maximized. Deleting the support vectors will change the position of the hyperplane. These are the points that help to build the SVM. To understand the concept of SVM, let's take an example. Consider the following image. You have a train dataset which consists of circle points and triangle points. From the sketch three lines, which one do you think is the best separator line? It is obvious that the best line is L_3. Also, there are many lines parallel with L_3. Again, which of these lines do you think is the better hyperplane? The best hyperplane (or in this example line) is the one that maximize the margins from the support vectors.

In the logistic regression, the output is the product of weights and input features which is a linear function and after that, squash the value within the range of [0,1] using the sigmoid function. If the squashed value is greater than a threshold value which is set to 0.5, its label is assigned to one, else the label is assigned to 0. In the SVM, the output of the linear function is taken and if this is greater than 1, the label of the new data will assign to one, for example blue points, and if the output is −1, the new label of new data will assign to another class, for example the red points. Since the threshold values are changed to 1 and −1 in SVM, we obtain this reinforcement range of values [−1,1] which acts as margin.

Now that you are familiar with the concept of SVM, let's take a look in training phase of SVM. From the above discussion, the SVM is choses the

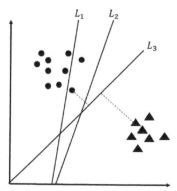

Figure 3.2. Example of three possible lines to separate the red and blue samples.

line that has the large margin to support vectors. Therefore, the training phase of SVM consist of a maximization optimization. In the SVM classifier, we are looking to maximize the margin between the data points and the hyperplane. The loss function that helps maximize the margin is hinge loss. The hinge loss is written as the following equation

$$cost(x, y, f(x)) = \begin{cases} 0 \, if \, yf(x) \geq 1 \\ 1 - yf(x) \, else \end{cases} \tag{79}$$

where in the above equation, x is input features, y is true label, and the $f(x)$ is the predicted label for input features. From the above equation, the cost is equal to 0 if the predicted value and the actual value are of the same sign. If they are in different signs, then the prediction is wrong and therefore, the loss value calculates for this input feature. Also, you can rewrite this cost function as the following equation:

$$cost(x, y, f(x)) = (1 - yf(x))_+ \tag{80}$$

You can also add a regularization term to the cost function. The objective of the regularization term is to balance the margin maximization and loss. The regularization term can avoid the SVM classifier to overfit. We will discuss more in detail about regularization and overfitting in the next chapter. After adding the regularization term, the cost function can be written as the following.

$$J = \min_{w} \sum_{i=1}^{n} (1 - y_i \langle x_i . w \rangle)_+ + \lambda \| w \|^2 \tag{81}$$

In the above optimization problem, n is the number data in train dataset and w is the parameters which specify the best hyperplane. In this equation, instead of using $f(x_i)$ the $\langle x_i . w \rangle$ term use. This due to the linear function that is used in SVM. This linear function has the projection interpretation. One the best optimization algorithms to solve this optimization, is gradient descent. In gradient descent, the weights are updated iteratively in the opposite direction of slop for each weight. You can write the update phase of gradient descent algorithm as the following

$$w = w - \alpha \frac{\partial J}{\partial w_i} \tag{82}$$

where α is the learning rate. If the learning is too small, the convergence takes too long and if it is large, it is possible to not converge. Therefore, it is important to choose the learning rate parameter wisely. To update the weights of SVM, you should compute the derivative of $\frac{\partial J}{\partial w_i}$. Now the loss function is defined, it is easy to take partial derivatives with respect to the weights to find the gradients. The partial derivatives of each of the loss function term with respect to the w_i is equal to the following equation:

$$\frac{\partial \lambda \|w\|^2}{\partial w_j} = 2\lambda w_j \tag{83}$$

$$\frac{\partial (1 - y_i \langle x_i.w \rangle)}{\partial w_j} = \begin{cases} 0 \ if \ y_i \langle x_i.w \rangle \\ -y_i x_{ij} \quad else \end{cases} \tag{84}$$

Therefore, the update stage can be rewritten as the following

$$w = \begin{cases} w - \alpha(2\lambda w - y_i x_i) \ if \ y_i \langle x_i.w \rangle \geq 1 \\ w - \alpha(2\lambda w) \qquad\qquad else \end{cases} \tag{85}$$

The first term in the above equation happens when there is a misclassification for the trained sample x_i. In other words, when the model makes a mistake on the prediction of the class of sample x_i, the update stage includes the hinge loss and regularization term. In contrast, when there is no misclassification for the train sample x_i. In other words, when the model correctly predicts the class of sample x_i, the update stage only include the regularization term.

Now, you are able to train a SVM classifier. But there are two other issues that we want to discuss about them. One issue you may notice is that the SVM classifier is a binary classifier and does not support multiclass classification natively. It only supports binary classification and separating data points into two classes. For multiclass classification, the same principle is utilized after breaking down the multiclass classification problem into multiple binary classification problems. The second issue is the kernel trick. As you may notice, the SVM classifier can separate only linear data points like shown in the above figure. For the nonlinear data point (or features), you can pass the input features from a kernel and map them to a new space which the data points can separate linearly. In the following we will focus on these issues.

There are some techniques to extend the binary SVM to multiclass SVM for the multiclass classification like cat, dog and pandas classification from images. The basic concept of all of those methods is that they use multiple binary SVM classifier. The first method is to map data points to a high dimensional space to gain mutual linear separation between every two classes. This is called a One-to-One approach, which breaks down the multiclass problem into multiple binary classification problems. In other words, in the One-to-One approach multi SVM classifier trains to separate each class from only one of the other classes. Therefore, in this method we need a binary classifier per each pair of classes. Another approach one can use is One-to-Rest. In that approach, the breakdown is set to a binary classifier per each class.

As we told before, a single SVM is a binary classifier and can differentiate between two classes. In nutshell, there are two approaches to classify data points from m classes' dataset:

1. One-to-Rest: In the One-to-Rest approach, the classifier can use m SVMs. Each SVM would predict membership in one of the m classes.

2. One-to-One: In the One-to-One approach, the classifier can $\dfrac{m(m-1)}{2}$ SVMs.

To be clearer about these approaches, let's take a multiclass classification example with 3 classes. In the following image, there are three green, red, and blue classes.

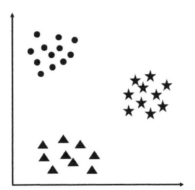

Figure 3.3. Example of a dataset with three class.

In the One-to-One approach, we need a hyperplane to separate between every two classes, neglecting the points of the third class. This means the separation takes into account only the points of the two classes in the current split. For example, in the above figure, four SVMs need. The first one, separates the red and blue points and tries to maximize the separation only between blue and red points. This classifier neglects the green points. The second one, find a line which is the best for red and green points without any attention to the blue points. Finally, the last classifier trains for green and blue points. In the One-to-Rest approach, we find a hyperplane to separate between a class and all others at once. This means that in contrast to the One-to-One approach, in this method the separation takes all points into account, dividing them into two groups; a group for the class points and a group for all other points. For example, the green line tries to maximize the separation between green points and all other points at once. The last approach is most popular technique to extend binary SVM classifier to multiclass SVM classifier.

Until now, we assume the features are linearly separable. This means that you can draw a straight line to separate the two classes, for example red and blue points in the first example. In practice, the data points usually are not linearly separable. The goal of kernel technique is that after the transformation to the higher dimensional space, the classes are linearly separable in this higher dimensional feature space. Afterward, you can then fit a decision boundary to separate the classes and make predictions. The decision boundary will be a hyperplane in this higher dimensional space. The "trick" is that kernel methods represent the data only through a set of pairwise similarity comparisons between the original data observations or train dataset with the original coordinates in the lower dimensional space, instead of explicitly applying the transformations $\phi(x)$ and representing the data by these transformed coordinates in the higher dimensional feature space. In kernel methods, the train dataset is represented by an $n \times n$ kernel matrix of pairwise similarity comparisons where the entries $(i.j)$ are defined by the kernel function $(k(xi.xj))$. This kernel function has a special mathematical property. The kernel function acts as a modified dot product. The kernel function accepts inputs in the original lower dimensional space and returns the dot product of the transformed vectors in the higher dimensional space. There are also theorems which guarantee the existence of such kernel functions under certain conditions. Let's take an example to understand the concept of the kernel tricks better. Consider some data points shown in following figure.

It is obvious that these data points cannot be separated with a linear decision boundary. However, the vectors are very clearly segregated and it looks as though it should be easy to separate them. The kernel techniques enable the SVM classifier to find a hyperplane to separate these data points. The main concepts of kernel tricks are to add a new dimension. In this example, you can add a third dimension. In this example, there are two dimension x and y. To create a new dimension like z, you should manipulate the x and y. For example, it is obvious from the above figure that the red and blue data points can separate with a circle decision boundary. Therefore, you

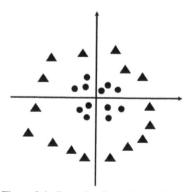

Figure 3.4. Example of a nonlinear dataset.

can create a new z dimension $z = x^2 + y^2$, which is the equation for a circle. This will give a three-dimensional space. If you draw the data points in the x and z dimensions, you can see that they are linearly separable. Now the SVM classifier can classifies these data points. This method is called kernel trick or technique. Unfortunately, in real world cases the features vectors are high dimensional and cannot plot them and decide which kernel function can map the data point to a linearly separable space. Therefore, it is hard to choose the best kernel function. In most cases, the best kernel functions are found by trial and error of desired kernel functions. In other words, you should train a SVM with a kernel function and validate it on the validation dataset. By using this process for all of your desired kernel functions, you can choose the kernel with the highest validation accuracy (or equivalently lowest validation error).

There are different kernels. The most popular kernel functions are the polynomial kernel and the radial basis function (RBF) kernel. Intuitively, the polynomial kernel looks not only at the given features of input samples to determine their similarity, but also combinations of these. If n is features vector dimension and d is degrees of polynomial, the polynomial kernel yields n^d expanded features. The mathematic formula of this kernel is written in the following

$$k(x,y) = (x^T y + 1)^d \qquad (86)$$

where the x^T is represent the transpose of the vector x. The RBF kernel is also called the Gaussian kernel. There is an infinite number of dimensions in the feature space because it can be expanded by the Taylor Series. The mathematic formula of this kernel is written in the following

$$k(x,y) = \exp -\gamma\|x - y\|^2. \quad \gamma > 0 \qquad (87)$$

where the γ parameter defines how much influence a single training example has [20].

Bayesian networks

A Bayesian network [48], belief network, also called directed acyclic graphical model, is a probabilistic graphical model. In this graph, a set of random variables and their conditional independence is represented with a directed acyclic graph (DAG). For example, a Bayesian network can be a graph showing the probabilistic relationships between diseases and symptoms. We can calculate the probabilities of the existence of different diseases given symptoms. Some practices implement inference and learning with a good performance. A dynamic Bayesian network is a Bayesian network that models sequences of variables, like speech signals or protein sequences. Influence diagrams a generalized Bayesian network that represents and solves uncertain decision-making problems.

Genetic algorithms

Genetic Algorithm (GA) [49] is a search-based optimization technique based on Genetics and Natural selection principles. It is usually applied for finding an optimal or near-optimal solution to complex problems. This problem normally would take a very long time, more than a lifetime, to solve. GA is frequently used to solve some kind of problems like optimization problems, research, and machine learning. GAs are search-based algorithms inspired by nature. It is based on the concepts of natural selection and genetics.

In GAs, there is a large number of possible solutions or a population to the given problem. These solutions, same as in natural genetics, go through recombination and mutation. As a result, it produces new children, and these steps are applied all over the generations. For each candidate solution or individual, based on the objective function, a fitness value is assigned. The most suitable individuals are given a higher chance to pair and yield more "fitter" individuals. Hence, we keep "evolving" better generations or individuals up to reach a stopping measure. This is in line with the Darwinian Theory of "Survival of the Fittest". In a random local search, we try several random solutions, following the best as yet. Genetic algorithms use historical information to achieve a much better result than random local search and are relatively randomized.

The genetic algorithm has some important advantages. The first one which is very important is that it does not require any derivative information. Computation of derivation is not practical in some practical problems. Also, the genetic algorithm is faster and more efficient as compared to the traditional methods. The other advantage of the genetic algorithm is that it has very good parallel capabilities. It is also useable for both continuous and discrete functions and also multi objective problems optimization problems. The genetic algorithm also provides a list of good solutions instead of a single solution. In addition, it always gets an answer to the problem, which gets better over the time. In general, the genetic algorithm is useful when the search space is very large and there are a large number of parameters involved. One of the main advantages of the genetic algorithm is that it can be a good solution for solving difficult problems. In computer science, there is a large set of problems, which are NP-Hard. What this essentially means is that, even the most powerful computing systems take a very long time to solve that problem. In such a scenario, GAs prove to be an efficient tool to provide usable near-optimal solutions in a short amount of time. Like any other optimization and learning algorithm, the genetic algorithm suffers from a few limitations. The genetic algorithms are not suited for all problems, especially problems which are simple and for which derivative information is available.

The computational amount of the genetic algorithm can be expensive for some problems which leads to a need for more time and resources.

Federated learning

Traditionally in machine learning, for training a model and to make a prediction, their data pipeline uses a central server instead of on-prem or the cloud, that hosts the trained model. In this architecture, all the sensors and devices collect the data and send it back to the server, and after processing the data, return it to the devices. This round-trip limits a model's ability for real-time learning. In other words, the standard machine learning approaches require centralizing the training data on one machine or in a datacenter. Consider models which want to train from user interaction like mobile or tablet devices. To train a model in this situation, the federated learning is introduced. Federated Learning enables devices like mobile phones to collaboratively learn a shared prediction model while keeping all the training data on the device.

Federated learning (FL) [50], in contrast, is an approach that downloads the current model and updates it using local data at the device itself (edge computing). These locally trained models are then aggregated at the central server, i.e., averaging weights. Then a single integrated and improved global model is sent back to the devices.

The overall procedure of the federated learning is similar to the following steps. In the first step, your device downloads the current model from the cloud storage. In the second step, the model is fine-tuned using the data on your mobile phone. In the third step, the changes in the fine-tuned model is summarized as a small focused update. Finally, this fine-tuned version of the model is sent to the cloud storage by leveraging the encrypted communication. In the cloud, the model is instantly averaged with other user fine-tuned models to improve the shared model. This procedure repeats when a user fine-tunes a new model. The important issue in the federated learning is that all the training data remains on your device, and the privacy of the user is protected. However, the individual updates are not stored in the cloud. The Google company currently test the federated learning in Gboard on Android, and the Google Keyboard. For example, when Gboard shows a suggestion query, the user's phone stores the information about the current context locally and which of the suggestions the user selects.

The main advantages of federated learning are lower latency, less power consumption, and ensuring the privacy of users. Also, the federated learning has another immediate benefit especially for the user. When the user provides an update to the shared model, the improved model on the user phone can be used immediately.

There are many theoretical and technical challenges which should be solved to make federated learning feasible. For example, in federated learning, the data is distributed over millions of devices around the world. However, these devices have significantly higher latency, lower throughput connections and intermittently available for training. While in the typical supervised learning algorithm, the optimization algorithm like mini batch gradient descent or stochastic gradient descent, optimizes on a dataset which is partitioned homogeneously across servers or a powerful computer.

CHAPTER 4
Some Practical Notes

|||||||||||||||||||||||||||||||||||||||||||||||||||||||||||||||||||||||||||||||||||||||||||||||||||||||||||||||||||||||||||||||||||||||||||||||||||||||||||||||||||

In the previous chapter, you learned about some of the important classifiers like artificial neural network, logistic regression, and support vector machine. When you train a classifier, usually you want to report the performance of your model in practice. Although, to have a good intuition of what happens when you train a classifier such as artificial neural networks, you should be familiar with concept of overfitting and underfitting. In this chapter, we will discuss how to split the dataset in training, validation and how to test in order to check the performance of a trained model. There are many metrics to measure the performance of the model. One of them is loss which we discussed in the neural network. In the reset of this chapter, we will introduce some of the important metrics to measure the performance of the model. Finally, we will introduce the ceiling analysis which is the great technique to debug a machine learning based system.

Resampling method

Resampling methods are essential techniques in the modern machine learning area. In these techniques, the training dataset is split into parts and a model of interest will fit on each part in order to obtain additional information. For example, in order to estimate the variability of a linear regression, the training dataset is split into subsets, to fit a linear regression to each new group, and then examine the extent to which the resulting different fits differ. Such an approach may allow us to obtain information that would not be available from fitting the model only once using the original training sample. Resampling approaches can be computationally expensive, because they involve fitting the same algorithm multiple times by using different subsets of the training data. However, due to recent advances in computing power, the computational requirements of resampling methods generally are not prohibitive. One of the most commonly used resampling techniques is cross validation which we discuss in this chapter.

Cross validation

Previously, the distinction between the test accuracy (or error) and the training accuracy (or error) was explained. The training accuracy is the average of the correct prediction of a model on the trained dataset. In general, we expect a high level of accuracy for the training dataset. The test accuracy is the average of the correct prediction of a model on the test dataset. Recall that model does not have any knowledge about the test data and are new observations for model. In contrast, we expect a lower test accuracy compared to trained accuracy. We will discuss more about test and trained accuracy. It is easy to compute the model performance on new observation if they are available, but unfortunately in most cases they are not available. The question is how to measure the accuracy of the model and report it to others? On simplistic approach is to report trained accuracy. As you may guess, this is not such a good idea. Because the model trains on a trained dataset and knows them and it is obvious that the model will have a high level of accuracy. Our goal is to report model performance in general and this is true when the model does not train on the test subset. The other approach is to split the dataset into two parts, trained and test subsets. In the training phase, model trains on training subset without any knowledge about the test subset. After the model trains, we use the test set to evaluate the performance of the model. Model accuracy in this technique is closer to actual model accuracy. Some questions may arise for this approach. The first one is how much of the tested accuracy is close to the actual accuracy of model? The second question is how to select the best hyperparameters? The answer for the first question is dependent on the test set. If this subset involves all distributions of observation, it is very close. To answer the second question, we need to evaluate the model performance for each hyperparameter. If we use the test set to select the best hyperparameter, we have a leakage between the test set and the train set which is wrong. In this situation, the model somehow learns for test set and now the test set is not valid in order to report the models performance. The best solution for this question is to use the validation set. Validation set is a subset of the training dataset which is not involved in the training phase, and only just to select the best hyperparameters and also avoid overfitting and underfitting. We will discuss in more detail about overfitting and underfitting. Note a few points, if we have a large dataset for example with a millions observation with much distribution, the dataset usually splits randomly into three trains (70%), validation (10%) and test (20%) subsets. In most cases, there isn't any very large dataset. For these datasets, splitting the dataset into three parts has three drawbacks. The first one is reducing the training samples which can yield into overfitting. The second one is the invalidity of the test accuracy because of a few numbers of new observations. Fortunately, there are some techniques for this situation too.

Leave one out cross validation

Leave one out cross validation (LOOCV) is very similar to the validation set approach, but it attempts to solve the expressed drawbacks. Like the validation set approach, leave one out cross validation splits the set of observations into two subsets. One subset is for validation and contains a single observation. The other subset is for training and consists of the remaining observations. The model will fit on the n − 1 training samples and validates on the validation sample which does not involve the training phase. The model accuracy for this validation set is a poor estimate for test accuracy because it is based upon just one new observation and is highly variable. To overcome this problem, we repeat this procedure by selecting the next sample as the validation subset and the remaining samples as the training set. We compute the validation accuracy for this validation subset too. Repeating this approach n times so that the model validates itself for all of the datasets. The final test accuracy of the model is the average of these n validation accuracy. This final test accuracy is a good estimation of test accuracy.

Leave one out cross validation has some major advantages over the previous approach. The first one is that it has less bias. This is because in leave one out cross validation we repeatedly fit the model using training sets which contain n − 1 samples and validation on the remaining sample. In leave one out cross validation each example is in the training set n − 1 times and in validation set 1 time. This is in contrast to the validation set approach, in which the training set is typically around 70% of the original dataset. The second advantage of, the leave one out cross validation is that it always has the same test accuracy. This is due to training and validation subsets which are not split randomly. In contrast, the previous approach will yield a different test accuracy when due to randomness in splitting train and validation subsets. Leave one out cross validation has the potential to be expensive to implement, since the model has to be fitted n times. This can be very time consuming if the n is large, and if each individual model is slow to fit. Thanks to the recent advantage on the power computation of computers, it is not very time consuming as in the previous decades. Leave one out cross validation is a very general method, and can be used with any kind of, predictive modeling. For example, we could use it with logistic regression.

K-fold cross validation

An alternative to leave one out cross validation is k-fold cross validation. This approach involves randomly dividing the set of observations into k groups, or folds, of an approximately equal size. The first fold is treated as a validation set, and the method is fitted on the remaining k − 1 folds. Then the validation accuracy is computed for the remaining group. This procedure is repeated k times; each time, a different group of observations is treated as a validation set. This process results in k estimates of the test accuracy. The k-fold cross validation estimate is computed by averaging these test accuracies. It is not hard to see that leave one out cross validation is a special case of k-fold cross validation in which k is set to equal n. In practice, one typically performs k-fold cross validation using k = 5 or k = 10. What is the advantage of using k = 5 or k = 10 rather than k = n? The most obvious advantage is computational. Leave one out cross validation requires fitting the statistical learning method n times. This has the potential to be computationally expensive. But cross validation is a very general approach that can be applied to almost any machine learning algorithms. Some algorithms like deep neural networks have computationally intensive fitting procedures, and so performing leave one out cross validation may pose computational problems, especially if n is extremely large. In contrast, performing 10-fold cross validation requires fitting the learning procedure only ten times, which may be much more feasible. There also can be other non-computational advantages to performing 5-fold or 10-fold cross validation, which involve the bias-variance trade-off.

Metrics

Before starting to explain the metrics used in machine learning areas, you should be familiar with concept of confusion matrix. Consider binary classification for whether an email is spam or not. There are four possible output which are True Positives (TP), True Negatives (TN), False Positives (FP) and False Negatives (FN). True positive and true negatives are the observations that are correctly predicted. Values of false positives and false negatives occur when your actual class contradicts with the predicted class. In the following, each of these four states are defined.

- True Positives (TP): These are the correctly predicted positive values which mean that the value of the actual class is yes and the value of the predicted class is also yes. For example, if the actual class value indicates that this email is spam and the predicted class tells you the same thing.

- True Negatives (TN): These are the correctly predicted negative values which mean that the value of the actual class is no and value of the predicted class is also no. For example, if the actual class says this email is not spam and the predicted class tells you the same thing.

- False Positives (FP): When the actual class is no and the predicted class is yes. For example, if the actual class says this email is not spam but the predicted class tells you that this email is spam.
- False Negatives (FN): When the actual class is yes but the predicted class is in no. For example, if the actual class says this email is spam but the predicted class tells you that this email is not spam

Now based on what you have learnt, we can explain commonly used metrics.

Accuracy

Accuracy is the most intuitive performance measure and it is simply a ratio of the correctly predicted observation to the total observations. In the following, the formula for accuracy is represented.

$$Accuracy = \frac{TP + TN}{TP + TN + FP + FN} \tag{1}$$

One may think that, if we have a level of high accuracy then our model is best. Yes, accuracy is a great measure but only when you have symmetric and balanced datasets. To be clearer, consider email spam, we have 95 spam samples and 5 non spam samples. Also, we have a model that always predict that an email is spam. The accuracy for this model is 95% which is great accuracy. Actually, the model has poor performance. But if we have 50 spam samples and 50 non spam samples, this model has 50% accuracy. Now you understand what a symmetric and balanced dataset means. There are other metrics to express the model performance for unbalanced test set.

Precision

Precision is the ratio of correctly predicted positive observations to the total predicted positive observations. The precision answers the question that of all spam emails that are labeled as spam, how many are actually spam? High precision relates to the low false positive rate. The formula for precision is expressed in the following:

$$Precision = \frac{TP}{TP + FP} \tag{2}$$

Recall

Recall or Sensitivity is the ratio of correctly predicted positive samples to the all samples in actual class – yes. The recall answers the question that for all spam emails that are actually spam, how many are predicted as spam?

High recall relates to the low false negatives rate. The formula for recall is expressed in the following:

$$Recall = \frac{TP}{TP + FN}$$ (3)

F1 score

Some models have high precision but low recall or vice versa. F1 Score is another metric that combines precision and recall. F1 Score is the weighted average of precision and recall. Therefore, this score takes both false positives and false negatives into account. Intuitively it is not as easy to understand as accuracy, but F1 is usually more useful than accuracy, especially if you have an uneven class distribution. Accuracy works best for balanced datasets. If the dataset is unbalanced, it's better to look at both Precision and Recall.

$$F1\, Score = \frac{2 \times Precision \times Recall}{Precision + Recall}$$ (4)

In industry and research after training a new model, you should report the performance of your model. The metric selection is depending on the case. It is important to know the problem and research other researchers or companies have done and how to report their performance and use which metrics. For example, in human action recognition, most datasets like UCF11 and UCF50 use accuracy to report model performance.

Normalization

Normalization is a technique that is often applied as a part of data preparation for machine learning. It is also called data preprocessing in the machine learning context. The goal of normalization is to change the values of features in the training dataset to be in a common scale, without distorting differences in the ranges of values or losing information. Normalization is also required for some algorithms to model the data correctly. Generally, each feature can vary differently. For example, assume that your training dataset has one feature with values ranging from 0 to 1, while the range of the other feature varies from 10 to 1000. The great difference in the scale of the numbers could cause problems when you attempt to combine the values as features during modeling. Also, in machine learning algorithms in which they have a training phase, this great difference in the scale of features can slow down the optimization and sometimes the model cannot converge. For example, consider the gradient descent in the training of a neural network. The presence of feature value in updating weights will affect the step size of the gradient descent. The difference in the ranges of features will cause different step sizes for each feature. To ensure that the gradient descent moves smoothly towards the minima and that the steps for gradient descent are updated at the same rate for all the features, it is better to scale the data before feeding it to the model. Therefore, it is necessary to preprocess the training data before training a model for them. Remember, in image problems like cat and dog classification, the input features are pixel value in which all of them are in the same range and therefore they do not need to be normalized. However, normalization avoids these problems by creating new values that maintain the general distribution and ratios in the source data, while keeping values within a scale applied across all numeric features used in the model. Some algorithms require that data be normalized before training a model. Other algorithms perform their own data scaling or normalization. Therefore, when you choose a machine learning algorithm to use in building a predictive model, be sure to review the data requirements of the algorithm before applying normalization to the training data. There are several methods that one can apply for normalization. The most popular and widely used techniques are rescaling (or min-max normalization) and Z-score normalization.

• Rescaling

The other name of this normalization is min-max normalization. It is the simplest normalization method. In this normalization method, the natural range of features maps into a standard range which is usually between 0 and 1. Sometimes, this range can be between –1 and 1. For example, the feature with range between 10 and 1000 and the other feature with range between 0 and 10 are map into a standard range between 0 and 1 (or between –1 and 1). To normalize the features by this method, you can use the following simple formula to scale to a range for each feature:

$$x' = \frac{x - x_{min}}{x_{max} - x_{min}} \tag{5}$$

Scaling to a range is a good choice when both of the following conditions are met. The first one is that you know the approximate upper and lower bounds on your data with few or no outliers. The second one is that your data is approximately uniformly distributed across that range.

• Z-Score normalization

Z-Score is a variation of scaling that represents the number of standard deviations away from the mean. You would use Z-Score normalization to ensure your feature distributions have to mean 0 and standard deviation 1. It is useful when there are a few outliers, but not so extreme that you need clipping. The formula for calculating the Z-Score of a point, x, is as follows:

$$x' = \frac{x - \mu}{\sigma} \tag{6}$$

where in the above equation μ is the population mean and σ is the population standard deviation. This normalization is also known as standardization. It is widely used in machine learning algorithms such as support vector machine and logistic regression.

Overfitting and underfitting

Consider that you have a model with poor performance. How do you find out the problem? Does the model train well? Does the model learn generalization concepts? In this section, we will discuss the generalization concept in machine learning and the problems of overfitting and underfitting. After this section, you will be able to figure out whether your model fits right or underfits or overfits.

Generalization refers to how well the model can predict the specific examples or data which it does not see yet. For example, consider that I show you an image of a cat and ask you to "classify" that image for me; assuming that you correctly identified as a cat, would you still be able to identify it as a cat if I just moved the cat three pixels to the left? What about if I turned it upside? Would you still be able to identify the cat if I replaced it with a cat from a different breed? The answer to all of these questions is almost yes because we as humans, generalize with incredible ease. Another good example is how a baby (our model) learns what is a cat or not. In the first stage, when he or she learns what a cat is, it can generalize it to all creatures which have four legs and label them as a cat. When their mother or father (supervisors) correct them, the baby's mind force itself to learn what actually a cat is. This procedure is called generalization. On the other hand, machine learning algorithms struggle very much to do any of these things; it is only effective in classifying that one specific image. While machine learning may be able to achieve superhuman performance in a certain field, the underlying algorithm will never be effective in any other field than the one it was explicitly created for because it has no ability to generalize outside of that domain. Generalization, in that sense, refers to the abstract feature of intelligence which allows us to be effective across thousands of disciplines at once. There is a terminology used in machine learning when we talk about how well a machine learning model learns and generalizes to new data, namely overfitting and underfitting. Overfitting and underfitting are the two biggest causes for the poor performance of machine learning algorithms. In the first one the model trains very accurately on a training dataset while in the latter one the model does not train well enough. In both cases the model lost its generalization ability.

In machine learning and statistics, the goodness of the fit states how well the model fits for a set of observations. Also, some machine learning researchers describe the goodness of the fit as a measurement which estimates how the approximation of the function matches (or model prediction) the target (or true label). Consider the underfitting problem. Underfitting occurs when a model is too simple which makes it inflexible in learning from the dataset. This simplicity comes from having too few features, selecting a simple model or too much regularization. Let's explain each of these factors separately. Consider the problem of predicting the house price in New York City. In this problem we use the linear regression model for house price prediction. If we only use the number of rooms and house area and ignore some important factors like the neighborhood, it is completely normal that our model be very simple and the boundary region would be very simple like a line. This caused by poor price prediction. The second factor is selecting a simple model. Again, consider the house price prediction problem. Also, consider that you have enough features and assume that you can fit a regression model with cubic polynomial. If you select a simple model like linear regression, your model can not learn from the training dataset. Also, it is normal that the model cannot generalize and has poor accuracy on the test or validation dataset. The third factor is high regularization. Some regularization techniques force the model to push the weights close to zero. If you use a cubic polynomial and high regularization, you force the weights w_1, w_2, and w_3 close to zero. In the other words, with high regularization you force the model to be like a quadratic or linear. Now like the previous factor, we have enough features but simple model which cannot learn from the training dataset and has a poor performance on the test dataset due to the low generalization concept.

Overfitting refers to the scenario where a model cannot generalize or fit well on the unseen dataset. A clear sign of overfitting is if its accuracy (error) on the testing or validation dataset is much less (greater) than the accuracy (error) on training dataset. In the other hand, high accuracy on a training dataset and low accuracy on the validation dataset indicates the overfitting problem. Overfitting is a term used in statistics that refers to a modeling error that occurs when a function corresponds too closely to a dataset. As a result, overfitting may fail to fit additional data, and this may affect the accuracy of predicting future observations. One may ask what is bad about a model that trains perfectly or on which the prediction on training dataset is very good? In almost all scenarios, the train dataset has noise. The term of noise refers to wrong labeling for a few observations, noisy features, and etc. Based on this knowledge, if a model learns perfectly on training dataset it means that it learns noises instead of the problem (for example house price). Now we can express that overfitting happens when a model learns the detail and noise in the training dataset to the extent that it negatively impacts the performance of the model on a new dataset. This means that the noise or random fluctuations in the training dataset is picked up and learned as concepts by the model. The problem is that these concepts do not apply to new datasets and negatively impacts the ability of the model to generalize. Now you can guess that underfitting is opposite of overfitting. Let's take an example to give more intuition about the difference between overfitting, underfitting and appropriate-fitting. Assume three students have prepared for a mathematics examination. The first student has only studied Addition mathematic operations and skipped other mathematics operations such as Subtraction, Division, Multiplication, etc. The second student has a particularly good memory. Thus, second student has memorized all the problems presented in the textbook. And the third student has studied all mathematical operations and is well prepared for the exam. In the exam student one will only be able to solve the questions related to Addition and will fail in problems or questions asked related to other mathematics operations. The first student is an example of underfits by the lack of knowledge. Student two will only be able to answer questions if they happened to appear in the textbook (as he has memorized it) and will not be able to answer any other questions. The second one overfits by remembering all problems without any generalization. Student three will be able to solve all the exam problems reasonably well. The last student is appropriate in terms of fits by learning the lessons correctly and gets the generalization intuition. In the following image, the underfitting, overfitting and appropriate-fitting problems are describing. Also, underfitting and overfitting is called high bias and high variance, respectively.

Until now, you learned what is overfitting and underfitting. Also, you can now explain how underfitting and overfitting cause the poor performance of the model. Now, we give you some simple techniques on how to understand

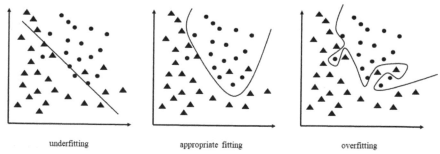

Figure 4.1. Example of underfitting (left image), appropriate fitting (center image), and overfitting (right image).

whether your model overfits or underfits or appropriate-fits. For detecting overfitting, underfitting or appropriate-fitting, we need training and validation datasets. In some references maybe instead of the validation dataset, they refer to a test dataset. In practice, the test dataset is not easy to access. Also, based on what we told them about validation and test subsets in the previous section, training and test subsets should not have any leakage. It is logical to use the validation dataset to check whether the model overfits, underfits or appropriate-fits. Because for detecting any kinds of fit, the training accuracy (or training error) and the validation accuracy (or validation error) are needed. If our model does much better on the training dataset than on the validation dataset, then we're likely overfitting. For example, the model performed with a 99% accuracy on the training dataset but only 50–55% accuracy on the validation dataset. It is overfitting the model and did not performed well on new observations. Also, we can figure out overfitting if the model performed with a 5% error on the training dataset but 15–20% error on the validation dataset. Conversely, if the model for train dataset has poor accuracy or high error, the model is likely underfitting. Finally, if the model does well on both the training and test datasets then we have the best fit (appropriate-fitting). For example, the model performed 90% accuracy on the training dataset and performs 88%–92% accuracy on the validation dataset. It is one of the best models [1, 12, 121].

By now, you are able to detect the underfitting and overfitting phenomena which is half of the way to debug the machine learning model, but it does not solve the problem. Fortunately, there are several techniques to solve these phenomena which is represented by machine learning researchers. Here are a few of the most popular techniques. In the underfitting case, it is obvious that by adding more data, the training accuracy (training error) will decrease (increase). Therefore, the remedy for Underfitting, is making new features, adding more complexity to your model, or reduce the regularization effect. In all solutions, the hypothesis space will expand. There are various techniques to prevent overfitting. A few of them are listed in the following.

1. Add more data

One probable reason for overfitting is the lack of enough data. Remember, it won't work every time, but training with more data can help algorithms generalize better. As the user feeds more training data into the model, it will be unable to overfit all the samples and will be forced to generalize. Users should continually collect more data as a way of increasing the accuracy of the model. However, this method is considered expensive and time consuming. Therefore, users should ensure that the data being used is relevant and clean. For adding more data, you can add some data and check whether the model training accuracy (training error) will increase (decrease) or not. If they move in the right way, you can add more data. Remember, if adding more data will decrease the effect of overfitting, but there are still other ways to reduce the overfitting effect.

2. Data augmentation

In the previous technique, we said that collecting more data will be expensive and time consuming. An alternative way to training with more data is data augmentation, which is less expensive compared to the former. If you are unable to collect more data, you can make the available datasets appear diverse. Let's clear the concept of data augmentation by an example. Consider the cat and dog classification problem. In our dataset, we have many images of cats and dogs. Consider one image of a cat. Will the nature of the cat change if the image moves a few pixels to right, left, down or up? The answer is negative. Therefore, we can move each image for a few pixels (say 10 pixels) to four directions. In this way, our dataset times by four. The other question one may ask is, will the nature of the cat change if add slight noise (like JPEG noise) to image? Of course adding slight noise does not change the nature of the cat. Data augmentation makes a data sample look slightly different every time it is processed by the model. The process makes each dataset appear unique to the model and prevents the model from learning the characteristics of the data sets.

3. Reduce complexity

Overfitting can occur due to the high complexity of a model, such that, even with a large dataset, the model still manages to overfit the training dataset. Reducing the Complexity method cause to decreasing the complexity of the model so that it makes it simple enough somehow the model can learn from the training dataset easily. Some of the actions that can be implemented include pruning a decision tree and reducing the number of parameters in Neural Networks like number of neurons in each layer and number of each layer. Simplifying the model can also make the model lighter and run faster. In this case, the model has less parameters to learn and more likely to does not overfits.

4. Remove features

Some models have built-in feature selection. For models that don't, you can manually remove irrelevant input features to improve generalization. Consider the house price problem. In this problem you have many features. Three of these features are height, width and area of the house. Are these three features being essential to train a good model? The answer is no. These three features are somehow related with each other. Therefore, instead of using three features of height, width and area, you can use only area of the house as the proper feature. This example clarifies the concept of the removing features. In practice, it is hard to understand which features are independent or dependent on others. Some algorithms like principal component analysis (PCA) are presented in order to extract more important and useful features.

5. Regularization

Regularization refers to a broad range of techniques for artificially forcing your model to be simpler. It is one of the most popular techniques to avoid overfitting. The method will depend on the type of machine learning algorithm. For example, in decision tree prune the tree is one of the regularization techniques whether in neural networks, dropout is one of the most commonly used regularization ways. Remember some of the regularization techniques are common for many machine learning models. Some of these regularization techniques, add an extra term to the cost function. Example of these techniques are L1 and L2 regularization. In contrast, some of these techniques are specific to the learning algorithm. Dropout and pruning the tree is two examples of these ways which uses in neural networks and decision tree, respectively. We will explain more about regularization in the next section.

6. Ensembling

Ensembles are machine learning methods for combining the predictions of multiple independent models. There are a few different methods for Ensembling. The most commonly used methods are Boosting and Bagging.

In boosting method some simple base models are used to increase their aggregate complexity of model. In this method, a large number of weak models are trained in a sequence such that each of them in the sequence, learns from the mistakes of the former model. Boosting combines all the base learners (which are weak learners) in the sequence so that bring out one strong and flexible model.

In contrast to boosting, the bagging method trains a large number of strong models (learners) which arranged in a parallel pattern. The final prediction is based on combining these models predictions to rich to a higher accuracy (or lower error). In other words, bagging combines all the strong models (learners) together to smooth out the final prediction. In the bagging method, the chance of overfitting complex models will decrease. If one or two models are overfit, the other models will compensate the poor predictions of these models.

7. Early stopping

Early stopping is one the popular techniques in neural networks which can help to prevent overfitting. When you are training a learning algorithm iteratively like neural network, you can easily measure how well each iteration of the model performs. These measurement criteria can be accuracy or cost on the validation dataset. The early stopping method, watches one of these criteria, for example the cost until the iteration for which the cost will start increasing.

In this iteration, the model loses the local or global minima and the cost increases. After that point, the ability of the model to generalize can weaken as it begins to overfit the training data. This point is that it is the best iteration to stop training the model.

Until now several techniques are presented to prevent overfitting. Some of these methods are mostly used in deep learning. For example, early stopping or dropout regularization layer are methods which are introduced to train better neural networks. However, the other methods like regularization are used in classical machine learning models. For example, in a SVM model you can add L1 or L2 regularization terms, or in decision tree model, you can reduce the maximum depth of the tree.

Regularization

As we said before, regularization is one of the most common methods to prevent overfitting. It is also probably one of the key concepts in the machine learning area. In machine learning problems, in most of the times many features are existed. For example, if in cat and dog classification problem each image is RGB and has 256 × 256 dimension, therefore, there are 196,608 features (number of image pixels). The key question for how this simple scenario may arise is are all features essential for our model? Some pixels in these images are describing the background for example the sky, yard or park. Do these pixels help us to train a better model? Some of these features (or pixels) can be useful but the others not. However, some pixels are representing the cat or dog in the image. In a nutshell, some of these pixels are important while some of others are not. Using these not essential features can lead the model to overfit. One way to remove these features is using feature selection methods like PCA. The second approach which is better is using regularization techniques. In general, regularization is the process to make things regular or acceptable. This is exactly why we use it for applied machine learning. In the context of machine learning, regularization is the process which regularizes or shrinks the weights towards zero. In any practical scenarios, the simple solution is better. There is a principle which states: "When faced with two equally good hypotheses, always choose the simpler". Therefore, regularization discourages learning a more complex or flexible model, to prevent overfitting. The basic idea of regularization is to penalize the complex models by shrinking the weights of model toward zero. In some regularization methods like L2 regularization, the weights of the features (pixels) which result in more loss to the cost function of the model, leading to a number close to zero. This strategy, reduces the effect of irrelevant features and cause it to reduce the model complexity. In other words, regularization penalizes the complex models by adding a complexity term that would give a bigger loss for complex models. To understand it, let's consider a simple relation for logistic regression. Mathematically, it is stated as below:

$$h = g(w_1 x_1 + w_2 x_2 + \cdots + w_n x_n + b) \tag{7}$$

where h is the probability of cat or dog in an image, w_i the weights of logistic regression model correspond to pixel x_i, and b is the bias term. Recall, in order to fit a model that accurately predicts the value of h, we require a loss function and optimized parameters, i.e., bias and weights. The loss function generally used for logistic regression is called the maximum likelihood and is represented in the formula in the previous chapter.

$$cost_{logistic} = \frac{1}{m} \sum_{i=1}^{m} -y^i * log(h(x^i)) - (1 - y^i) * log(1 - h(x^i)) \tag{8}$$

The above cost function is the logistic regression without regularization term. The model will learn the cat and dog problem by the means of this loss function. Based on the training dataset, the weights will adjust. If the dataset is noisy, it will face overfitting problems and estimated weights won't generalize on the unseen data. Also, if the dataset has some irrelevant feature (pixels), the model complexity will increase and overfitting occurs. This is where regularization comes into action. It regularizes these learned estimates towards zero by penalizing the magnitude of the weights. There are two main regularization techniques, namely Ridge Regression (L2 regularization) and Lasso Regression (L1 regularization). They both differ in the way they assign a penalty to the weights.

Ridge regression

This regularization technique performs L2 regularization and it also called L2 regularization. It modifies the logistic regression by adding the penalty (shrinkage quantity) equivalent to the square of the magnitude of weights. The new cost function for logistic regression using L2 regularization represents by the following formula.

$$cost = cost_{logistic} + \alpha \sum_{i=1}^{m} w_i^2 \tag{9}$$

By using the new cost, the weights are estimated using this modified loss function. In the above equation, the parameter α along with shrinkage quantity is the hyperparameter of L2 regularization and also called a tuning parameter that decides how much we want to penalize our model. In other terms, tuning parameter balances the amount of emphasis given to minimizing the logistic regression cost function vs minimizing the sum of the square of weights. The α parameter can play a key role to affect the weights and how much the logistic regression will be complex. Before explore the effect of α parameter, let's get more intuition about the regularization term. All machine learning problems can have many solutions. For example, in the cat and dog problem, there are many models that have the best performance. Based on all these models, the models that have lower weights are more acceptable. Also, large values of weights can extend the search area of the optimization problem. Consider the ith feature (or pixel) of the cat and dog classification problem. The L2 regularization term in the above equation will force the ith weight to be small by penalizing the model and add the square of the ith weight. Therefore, the appropriate-fit will tend to have a low value of weights.

Now, let's see how the value of α alpha affects the estimates produced by L2 regularization. When α is equal to zero, the penalty term has no effect. It means it is only the logistic regression loss function. We will expect to get the same weights as simple logistic regression. In contrast, when α is tend to infinity, the L2 regularization force the weights to have values close to zero because the modified loss function will ignore the core loss function and minimize weights square and eventually end up taking the parameter's value close to zero. When α is between zero and infinity, the weights of logistic regression will be somewhere between 0 and 1. This is due to the fact that the square of a number between 0 and 1 is less than the number. Therefore, if the weights are between 0 and 1, the extra loss will be at their minimum states. That is the reason for selecting a good value of α is critical.

Also the good effect of L2 regularization to prevent overfitting, this method has two drawbacks. The first one, the L2 regularization does not helps in Feature Selection. It decreases the complexity of a model but does not reduce the number of independent variables since it never leads to weights being zero rather than only minimizing it. Therefore, this method is not suitable for feature selection. The second one is model interpretability since it will shrink the weights for the least important predictors, very close to zero but note it will never make them exactly zero.

Lasso regression

Another regularization method is Lasso Regression. This regularization technique performs L1 regularization. Therefore, it is also called L1 regularization. Again, consider the logistic regression problem. It modifies the logistic regression cost by adding the penalty equivalent to the sum of the absolute value of weights. In terms of math it can be written as the following:

$$cost = cost_{logistic} + \alpha \sum_{i=1}^{m} |w_i| \qquad (10)$$

By using the new cost, the weights are estimated using this modified loss function. Again, in the above equation, the parameter α along with shrinkage quantity is hyperparameter of L1 regularization and is also called a tuning parameter that decides how much we want to penalize our model. L1 regularization is different from L2 regularization as it uses absolute weight values for normalization. As loss function only considers the absolute weights, the optimization algorithm (like gradient descent) will penalize high weights.

Like the L2 regularization, let's see how the value of α alpha affects the estimates produced by L1 regularization. When α is equal to zero, the penalty term has no effect. It means it is only the logistic regression loss function. We will expect to get the same weights as simple logistic regression. In contrast, when α is tend to infinity, the L1 regularization force the weights to have values close to zero because the modified loss function will ignore the core loss function and minimize absolute weights and eventually end up taking the parameter's value close to zero. Again, when α is between zero and infinity, the weights of logistic regression will be somewhere between 0 and 1.

In L1 regularization, the penalty term has the effect of forcing some of the weight estimates to be exactly equal to zero. This means there is a complete removal of some of the features for model evaluation when the tuning parameter α is sufficiently large. Therefore, the L1 regularization method also performs Feature selection and results sparse models.

The L1 and L2 regularization methods are very similar. In the L2 regularization, the weights penalize with squares of them while in L1 regularization with their absolute values. Both of L1 and L2 regularization are mostly used to reduce the overfitting in the model, but, the L1 regularization also has a feature selection behavior.

Dropout regularization

Dropout is another regularization method which is most common in deep neural networks. In deep learning literature, they consider dropout regularization as a layer. In the Dropout method, a large number of parallel neural networks with different architectures train on training dataset. During training procedures, some neurons in desired layers are randomly ignored or dropped out. In other words, dropout will temporarily remove a unit out from the network, along with all its incoming and outgoing connections. This technique reduces the number of neurons in each iteration, and therefore, reduces the number of trainable parameters and leads to simplifying the neural network model. Removing some neurons randomly in some layers, also cause it to train large neural networks with different architectures in each iteration. For example, if there is a neural network with 5 layer and the dropout rate is 50%, in 10 iterations it is probable to train 10 different neural networks. In the training phase in neural networks, units may change somehow that fix up the mistakes of the other units. This may lead to complex co-adaptations. This can result in overfitting because these co-adaptations do not generalize to unseen data. Dropout has the effect of making the training process noisy, forcing nodes within a layer to probabilistically take on more or less responsibility for the inputs. This is because of co-adaption of neural network layers to correct mistakes from prior layers. This technique can help breaks this situation and the model will be more robust after training. A side effect of dropout regularization is to encourage the neural network model to learn a sparse representation. Therefore, it can be used as an alternative regularization method for encouraging sparse representations in autoencoder models. The dropout has the effect of reducing the capacity of the neural network during training since, the outputs of a layer under the dropout are randomly subsampled.

You are now familiar the concept of dropout regularization. Now it is the time to learn how to apply this regularization method to the neural network. As we discussed before, the dropout regularization can apply to each layer of neural network such as dense fully connected layers, convolutional layers, and recurrent layers like the long short term memory network layer. Therefore, each layer of neural network can have dropout regularization or not. Remember, it is not logical to use dropout in the first layer, since the raw input features ignore randomly. Also, it is not used on the output layer. While each layer in neural networks can have dropout regularization, the number or percentage of dropping out units (or neurons) for each layer must be specified. Therefore, the dropout regularization method has a hyperparameter that specifies the probability in which outputs of the layer drop out. You can interpret the inverse statement of this fact in which the dropout hyperparameter specifies the probability of the outputs of the layer that are retained. For example, if a layer in neural network has 100 neurons (or units) and the dropout probability is 50% (or equivalently 0.5), during training, in each iteration only the 50 neurons will remain and the other 50 neurons will ignore. It is obvious that if this probability is set to 1, the behavior of the layer is like when there is not any dropout regularization. The interpretation of the dropout hyperparameter is the probability of training a given node in a layer, where 1.0 means no dropout, and 0.0 means no outputs from the layer. A common value for the dropout probability of hidden layers is 0.5 and a value close to 1.0, such as 0.8, for the last and first layers. In most of the modern deep networks, such as ResNet and VGG, the first layers and last layers have less neurons compare to middle layers. Therefore, these middle layers have more complexity and are more probable to cause in overfitting. In contrast, the first and last layers have less neurons and have more important features for prediction. Therefore, it is logical to set high probability for the first and last layers to keep most of the neurons in training and less probability for the middle layers which create more meaningful features.

Until now, everything we discuss is for the training phase. What should we do in test stage? If we use dropout regularization in the test phase, the output prediction for an observation will be random each time which is not true. Dropout is only used during training and not used after training when making a prediction with the fit network. Now, the weights of the network will be larger than normal because of dropout. Therefore, before finalizing the network, the weights are first scaled by the chosen dropout rate. In other words, the output of activation of each neuron during training will scaled by the dropout probability. In the test stage, we only use scaling procedure to normalize the output of each neuron. For example, if a neuron is retained with probability p during training, the outgoing weights of that unit are multiplied by p at test time. The dropout regularization can be used with all types of neural network models like Multilayer Perceptrons, Convolutional Neural Networks, and Long Short-Term Memory Recurrent Neural Networks. The famous deep learning libraries like Tensorflow and Pytorch have dropout layer.

It is common for larger networks (more layers or more nodes) to more easily overfit the training data. When using dropout regularization, it is possible to use larger networks with less risk of overfitting. There is a relationship between dropout probability and number of neurons. A good rule of thumb is to divide the number of neurons in the layer before dropout by the proposed dropout rate and use that as the number of nodes in the new network that uses dropout. For example, a layer of neural network with 100 nodes and a proposed dropout rate of 0.5 will require 200 nodes when using dropout [1].

Now you have learnt some important regularization techniques such as L1 regularization, L2 regularization and dropout. Also, there are other regularization method which are out of the scope of this book.

Ceiling analysis

Until now, you learnt how to find out your model problems during training and how to avoid the undesirable phenomena like overfitting and underfitting. In this section, we will introduce another technique to debug your machine learning system. This method is called ceiling analysis. Ceiling Analysis is a technique to systematically find the weakest component of your system and therefore, improve your system. Also, ceiling analysis is able us to find out which component of our system with improvement can increase the overall performance of the system. This technique can save our precious time and resources. Ceiling analysis is the process of manually overriding each component in your system to provide 100% accurate predictions with that component. Thereafter, you can observe the overall improvement of your machine learning system component by component. To be more clear, let's take an example. Suppose, you are a machine learning engineer and you have limited time and resources like GPU and RAM. Your boss, defines a new project to propose a new system which will fetch the images in Instagram and classify them into three categories. The first category is images with one male face and without any text in the image. The second category is images with one female face and without any text in the image. The Third category is all the other images. Your overall system pipeline will look like the following. The first stage is getting images from Instagram, the second stage is text detection, the third stage is face detection, and the last stage is gender detection. Consider the first stage which is getting images from Instagram and this works perfectly. Therefore, there are three machine learning model in your system. The first one is text detection model which identifies whether there is text in an image or not. The second one is face detection model which detect faces in an image and check whether only one face in the image or not. The third one is gender detection model which classifies the gender of detected face in the image. All these model can be a Convolutional Neural Network or any other machine learning algorithms. For test, there are 10,000 images that it can get from Instagram. Assume that the accuracy of your system is 68%. This is not a good accuracy for practical solutions. You now must find a solution to solve this problem as soon as possible and also, you do not have enough resources. You may think the cause of poor performance of your model is text detection. Maybe it is from face detection or also it can result of poor gender detection. How can you find which model will have the most effect in your model? One naïve way is to train new models which is not practical because of the time and resource limitation. The ceiling analysis technique can help you to debug your model and improve the performance of your system without collection of more datasets and training new models. Consider this accuracy as a baseline accuracy. In ceiling analysis, the component manually overwritten to provide 100% accuracy. For example, you can detect text in 10,000 test images. This yields 100% accuracy in text detection model. This procedure should work chronologically until all our components are manually overridden, and observe the changes in accuracy, one component at a time. For example, after manually detecting text in images, you detect faces in images manually and after that classify the gender of each face by yourself. By the end of this process, your algorithms and overall system will be predicting 100% accuracy. Based on what we said, you are now detecting text in images where the text detection has 100% accuracy and running your system to observe how much the accuracy will increase. After that, you detect text and faces in images where text and face detection have 100% accuracy and run your system to observe how much the accuracy will increase. And in the final stage, you detect text, faces and genders manually which all components have 100% accuracy and run your system to observe how much the accuracy will increase. It is obvious that in the last stage the system accuracy reach to 100%. After doing this procedure, your results are as follows:

Baseline accuracy: 68%

Perfect text detection: 69%

Perfect face detection: 78%

Perfect gender detection: 100%

As you can see, the perfect text detection improves the accuracy of the model only by 1%. This suggests you do not invest your time and resources to improving the text detection model. It also tells you that probably your text detection model is good enough. In a nutshell, improving the text detection model is not the issue here. Moving onto the face detection, the face detector model looks quite strong. Also, the improvement of this model has 9% effect which is quite worth it to invest time in when you will have more free time, but not now. The final improvement is the gender recognition model. It appears that gender recognition is struggling the most, yielding a 22% improvement in accuracy with perfect results. Now, this is by far the weakest link in your overall system and by improving this model, your baseline accuracy ideally can reach to 90%. Ceiling analysis suggests that working on the gender recognition model of your system will yield the best overall improvements. Now, you debug your system and find out how to invest your time and resources to train a new gender recognition model.

In abstract, the ceiling analysis method debugs your system and can guarantee that you are focusing on the correct component of your system of which improvements are likely to yield substantial improvements in accuracy. As a project manager, you do not want to allocate resources and time to tasks that will not yield results and be confident that what you assign your team to is worthwhile. Remember, the ceiling analysis method is not just for machine learning area. It is a method that can help you think correctly.

CHAPTER 5

Deep Learning

IIIIIIIIIIIIIIIIIIIIIIIIIIIIIIIIIIIIIIIIIIIIIIIIIIIIIIIIIIIIIIIIIIIIIIIIIIIIIIIIIIIIIIIIIIIIIIIIIIIIIIIIIIIIIIIIIIIIIIIIIIIIIIIIIIIIIIIIIIIIIIIIIIIIIIIIIIIIIIIIIIIIIII

Deep Learning (DL) is considered to be a branch of machine learning algorithms which are based on artificial neural networks. DL algorithms also can be categorized into supervised, semi-supervised, or unsupervised learning [1, 51, 52]. DL consists of a wide variety of structures, including, but not limited to, deep neural networks, recurrent neural networks, and convolutional neural networks, which have been utilized in different applications such as computer vision [53], speech recognition [54], fingerprint-based localization [59], filtering of social network contents [55], bioinformatics [56], analyses of medical images [57], and game applications [58]. The DL algorithms in real-world applications result in great performance even when compared with humans.

The term "neural networks" has been derived from human brain functionality [9]. In the human brain, a neuron assembles the signals of the other neurons through dendrites. The neuron releases the electrical signals from the axon, which is divided into many branches. Then, a synapse changes the signal of the axon to electrical effects. If a neuron gets an adequately huge input, it sends the electrical signal from the axon. The learning process happens when the efficacy of the synapses changes. The behavior of neurons in human brain functionality can be considered in order to study the nature of neuronal cells. Therefore, there are still many fascinating efforts to follow the brains learning paradigm by building new networks based on artificial neurons.

Artificial neural networks usually consist of numerous densely connected units that are considered neurons of the model. The matrix weights model the synapse. Most ANNs do not model the structure of the dendrites and axons. Each unit transforms the input signals and sends them to other units. This process consists of two steps: (1) the input signals are multiplied by the associated weights and are added to all weighted signals. (2) the unit utilizes an activation function to convert the weighted signals to output.

The general behavior of an ANN relies on both the weight matrices and activation functions that are used for the units. Many functions can be used as activation functions, including but not limited to linear, relu, elu, tanh, softmax, and sigmoid. Each activation function has special properties and is used for particular tasks. For example, the sigmoid activation function is usually utilized for binary classification problems in the output layer. As another example, the softmax activation function is usually utilized in the output layer for multi-class classification problems.

For doing a particular task, the associated weights of units must be set appropriately through optimization. The associated weights indicate the possibility of affection by other units. In other words, the weight matrices determine the power of the affections. There is a categorization for the units in the literature, whether they belong to input, output, or intermediate layers. The intermediate units usually are called hidden units, which connect the input layer to the output layer. The intermediate layers (e.g., number of units, number of layers, type of units, and so on) can be easily modified compared with the input and output layer. The interconnection weights between layers determine whether a unit is active or not. Therefore, the optimization of layers is an important step in assigning the weights for the hidden units. Researchers have introduced many optimization algorithms and learning paradigms for assigning the weights of neural networks. However, there is no specific and global rule for achieving all of the given tasks. It means there are still many challenges to simulate the human brain functionality.

The commonly used term "deep" in deep learning alludes to multi-layer units in the network instead of using a single hidden layer. Recent research shows that linear activation function cannot be used for advanced tasks in an ANN. Due to the nonlinear operations in the human brain, it is more practical to use other activation functions with nonlinear behaviors. In the previous years, the width of units was fixed, and the weights of units were only updated during the optimization process. However, nowadays, the width or parameters of the activation functions can also be learned and optimized during the training phase [120]. The performance of neural networks with adaptable activation functions usually shows a better performance in terms of accuracy and robustness.

Overview

The current deep learning frameworks are built based on artificial neural networks, particularly convolutional neural networks (CNNs), which show great performance in image classification problems. These neural networks can also be a designative formula or hidden variables of a generative algorithm such as Boltzmann machines or deep belief networks [60].

In deep learning, the layers play a key role in converting the input data to the final output. Each layer trains for converting its input to other representation, which are also called features. In image processing, the input data can be a 2D/3D array or vector of pixels. For the array-like inputs, convolutional layers are usually used in the input, and hidden layers and dense layers are used for the output layer to find the probability of belonging with the input to each class. Considering the face of humans in a classification problem, the output of the first hidden layer maybe be the faces' edges, the second layer may be the features of faces such as eyes, eye-lash, nose, ear, and so on. The output layer represents the probability based on the detailed representation of each previous layer. Note that there is no special rule to choose the number of layers and neurons for each layer.

As mentioned before, the word "deep" refers to the number of layers (usually more than two hidden layers). Especially, deep learning algorithms which belong to a notable depth of credit assignment path (CAP). The term CAP refers to a set of chain processes, which converts the input data to the outputs and indicates the possibility of a usual interconnection between input and output. For typical neural networks, CAP's depth is set based on the number of hidden layers. It equals the number of hidden layers plus one (because the final layer, called output layer, is also parametrized like the hidden layers). In recurrent neural networks, since the was signal turned around into a single layer many times, the CAP depth can be unlimited [61]. There is no special definition for the term "deep" in deep learning; however, it is called that when the CAP depth is more than two. A CAP with a depth of higher than 2 has indicated a great performance in simulations and can be used to chase any function. Increasing the number of layers usually shows better performance till reaching a certain layer. After that certain layer, adding more layers could not help to enhance the performance and only raise the training time.

DL structures can be built based on greedy layer algorithms and can be utilized for many tasks such as supervised, unsupervised, and semi-supervised algorithms. For supervised learning problems, DL algorithms remove the engineering and human efforts for doing special tasks on the data by transforming the data to the hidden layers and extracting features. DL algorithms can also be utilized for unsupervised learning problems when the labeled data is not available, or the labeled data is not sufficient. For instance, autoencoders [62] and deep belief networks [63] can be used for unsupervised learning problems. On the other hand, there is a new deep learning-based algorithm, called generative adversarial networks, which can generate synthetic data based on a set of limited data samples. The distribution of generated data is similar to the real data, but they are not the same.

Interpretations

Deep Learning is usually interpreted based on universal approximation (UAT) [64] or probabilistic theorem (PT) [65]. The typical UAT is one that considers the capability of feedforward neural networks using a hidden layer with a finite size to estimate a continuous function. The UAT for deep neural networks considers the capability of networks using a limited width; however, the depth of neural networks can be grown. Considering the width of the deep neural network using the relu activation function for the units, it can estimate any integrable function if the width is larger than the input dimension. On the other side, considering the smaller or equal width for the neural network compared with the input dimension, the deep neural network is not a global approximator [66].

The PT stems from the machine learning concept and shows the connection and optimization of training and testing procedures, such as fitting the machine learning algorithm to the data and generalization. More generally, the PT takes the nonlinearity of activation into account as a cumulative distribution function (CDF). The probabilistic explanation has led to the invention of dropout layers of neural networks to avoid the overfitting problem.

Artificial neural networks

As mentioned, ANNs are built based on human brain functionality, with neuron units interconnected like the world wide web. The human brain consists of thousands of billion cells named neurons. A neuron is built by a cell body responsible for carrying the data between the inputs and outputs of the human brain.

An ANN has thousands of billion of neurons, which are associated with units. They are built with inputs and outputs. The input nodes (units) collect different data based on the inter-weight system, and the ANN tries to learn the information of data based on inputs and outputs. It works based on the human brain functionality and requires special rules and guidelines such as labels and outputs. It also needs an algorithm to modify the weights based on an optimization procedure. For instance, the backpropagation algorithm is a well-known method for the optimization process that consists of two steps: (1) forward propagation (2) backward propagation.

An artificial neural network in supervised learning tasks consists of two phases called training and test phases. In the training phase by studying the labeled data, it learns the pattern of data, which converts the input data to the available labels. During the training phase, an optimization process is performed to minimize a cost function. For a regression problem, mean squared error (MSE) is usually utilized as a cost function defined and based on subtracting the actual outputs and estimated outputs. The cost function is minimized, the weights are updated.

An ANN in a binary classification problem learns how to convert the input to yes/no outputs during the training procedure. The binary classification problem is usually used in DL. It is a simple and important task that the used model attempts to classify the input features to a binary answer (Yes or No). For example, we can classify whether the input image is a fruit or not by giving the model a large data set during the training phase. We can give the model two sets of images: (1) One set of images can be the fruit images such as apple and oranges, which we mark them as 1 (it means Yes) (2) other sets of images can be anything except the fruit images, and we mark them as 0 (it means no). In the training phase, the ANN model learns the features and patterns of fruits such that in the test phase, it can recognize the fruit images with high accuracy.

Deep neural networks

Based on the definition, a deep neural network (DNN) is an ANN with several hidden layers that connect the input layer to the output layer [67]. Although different neural networks exist in literature, they are created based on the same instruments such as biases, weights, units, and activation functions. The mentioned instruments work based on human brain behavior. The DL algorithms can be trained analogous to other ML algorithms. For instance, if the goal of a DL model is to discern the chickens from other images, a collection of images which consists of chickens with label 1 is given to the model and trained based on training data.

DNNs can approximate complex non-linear functions. DNN models convert the input features to the detailed features of inputs, and then the estimation is performed based on the detailed features in the latest layer [68]. The hidden layers are in charge of detailed features. The hidden layers near the input layer consist of high features, and the layers near the output layer consist of more detailed features [67]. For example, the sparse multivariate polynomials are easy to estimate with deep neural networks when compared to shallow networks [69].

Although deep learning architectures consist of different types, they have a few basic architectural features. For example, generative adversarial networks (GANs) which were introduced in 2014 have many variants and evolved versions; however, all of them are inspired by the main model of GAN. Every architecture is useful in a particular field of study. It is not always easy to compare different architectures together because they have been designed for a special task. The DNN weights are first are initialized with random values; however, these random values do not give a good performance, and they must be optimized and updated based on training data. After the training phase, the deep model can recognize the pattern of new input data called test data.

Deep learning algorithms

DL algorithms refer to a set of various algorithms. Although there is no perfect algorithm to do all possible tasks, some algorithms have better performance to do some particular tasks. Therefore, choosing the right algorithm forces us to know the underlying and the principles of the algorithms.

Convolutional neural networks

The convolutional neural network, or CNN for short [75], is a specialized type of neural network model designed for working with two-dimensional image data, although they can be used with one-dimensional and three-dimensional data. Convolutional Neural Networks are very similar to ordinary Neural Networks which we discussed in the previous chapter. The convolutional neural networks consist of neurons that have trainable weights and biases. Similar to the neural networks, each neuron receives some inputs, performs a dot product and optionally follows it with a nonlinearity like sigmoid, ReLU or Tanh. The whole network still expresses a single differentiable score function from the raw input like image pixels on one end to class scores at the other. The convolutional neural networks like the ordinary neural networks have a loss function on the last fully connected later. In nutshell, we can use all techniques designed in neural networks to develop an efficient convolutional neural network. One may ask what changes in convolutional neural networks? The convolutional neural network architecture make the explicit assumption that the inputs are images, which allows us to encode certain properties into the architecture. While in the neural network the input is one dimensional data. These networks make the forward function more efficient to implement and vastly reduce the amount of parameters in the network.

As you recall from the previous chapter, in ordinary circumstances neural networks receive an input which is a single vector, and transform it through a series of hidden layers. Each hidden layer is made up of a set of neurons, where each neuron is fully connected to all neurons in the previous layer, and where neurons in a single layer function completely and independently and do not share any connections. The last fully connected layer is called the output layer and in classification settings it represents the class scores or probabilities. Although the neural networks are powerful in many problems, they do not scale well for images. Consider CIFAR-10 dataset which is one of the most popular datasets in image recognition context. In this dataset, images have the same size which is $32 \times 32 \times 3$ which the width, heights and color channels is 32, 32, and 3 respectively. Therefore, a single fully connected neuron in the first hidden layer of a regular neural network have $32 \times 32 \times 3$ weights which is 3072. This amount of trainable parameters still seems manageable, but clearly this fully connected structure does not scale for images with high dimensions. For example, consider a low quality image with size $200 \times 200 \times 3$. The first hidden layer for these images lead to neurons that have $200 \times 200 \times 3$ which is 120,000 weights. Moreover, to build a powerful neural network, you certainly want to have several layers with huge parameters, and therefore the number of trainable parameters increases exponentially. Clearly, this full connectivity is wasteful and the huge number of parameters would quickly lead to overfitting.

Convolutional neural networks take advantage of the fact that the input consists of images and they constrain the architecture in a more sensible way. In particular, unlike a regular neural network, the layers of a convolutional neural network have neurons arranged in 3 dimensions which are width, height, depth. Note that the word depth refers to the third dimension of an activation volume, not to the depth of a full Neural Network, which can refer to the total number of layers in a network. For example, the input images in a CIFAR-10 dataset are an input volume of activations, and the volume has dimensions $32 \times 32 \times 3$. The neurons in a convolution layer will only be connected to a small region of the layer before it, instead of all of the neurons in a fully connected manner. In the CIFAR-10 has 10 classes and therefore the final output layer for this dataset has dimensions $1 \times 1 \times 10$, because by the end of the convolutional neural network architecture the full image will be reduced into a single vector of class scores, arranged along the depth dimension. Remember that every layer of a convolution layer transforms the 3 dimensional (or equivalently 3D) input volume to a 3D output volume of neuron activations.

A simple convolutional neural network is a sequence of layers, and every layer of a convolutional neural network transforms one volume of activations to another through a differentiable function. In general, there are three main types of layers to build convolutional neural network architectures. These layers are a convolutional layer, pooling layer, and fully connected layer. The first two layers are new but the fully connected layer is exactly the same in the regular neural networks context. To build a convolutional neural network, these layers will be stacked to form a full CNN. For example, in a simple schematic convolutional neural network for CIFAR-10 dataset can have the following architecture: [input layer $\rightarrow$ convolution layer $\rightarrow$ ReLU activation layer $\rightarrow$ pooling layer $\rightarrow$ fully connected layer]. The input layer in this network is INPUT $32 \times 32 \times 3$ and will hold the raw pixel values of the image, in this case an image of width 32, height 32, and with three color channels Red, Green, and Blue. The second layer is the convolution layer which will compute the output of neurons that are connected to local regions in the input, each computing a dot product between their weights and a small region in which they are connected to in the input volume. If there are 12 filters in convolution layer, the output of this layer is a volume with size $32 \times 32 \times 12$. The third layer is ReLU activation layer which will apply an elementwise ReLU activation function (which is the $max(0.x)$). This leaves the size of the volume unchanged which in the our example is $32 \times 32 \times 12$. The fourth layer is the pooling layer which will perform a downsampling operation along the spatial dimensions. The spatial dimensions mean the width and height of the input volume to this layer. In our example, if the downsampling performs on each 2×2 region, the resulting output is a volume with size of $16 \times 16 \times 12$. The fifth and last layer is a fully connected layer which will compute the class scores. In our example, the output of this layer is a volume of size $1 \times 1 \times 10$. Remember that in CIFAR-10 there are 10 classes and therefore each of the 10 numbers in the output of the fully connected layer correspond to a class score. Like the neural networks, each neuron in this layer will be connected to all the numbers in the previous volume.

In general, the convolutional neural networks transform the original image layer by layer from the original pixel values to the final class scores. Note that some layers contain trainable parameters while some of the other layers does not have trainable parameters. In particular, the convolution and fully connected layers perform transformations that are a function of not only the activations in the input volume, but also of the parameters which are the weights and biases of the neurons. On the other hand, the ReLU and pooling layers will implement a fixed function. The parameters in the convolution and fully connected layers will be trained with gradient descent so that the class scores that the convolutional neural network computes are consistent with the labels in the training set for each image. In summary, a convolutional neural network architecture is in the simplest case that a list of layers that transforms the image volume into an output volume. There are a few distinct types of layers like convolution, ReLU, pooling, and fully connected layers. Each Layer accepts a 3D input volume and transforms it to a 3D output volume through a differentiable function. Each layer may or may not have parameters. For example, the convolution and fully connected layers have trainable parameters while the ReLU and pooling layers does not have trainable parameters. Each layer may or may not have additional hyperparameters. For example, the convolution, fully connected, and pooling layers have hyperparameters while the ReLU layer does not have hyperparameters. Now, we will describe the most popular layers in convolutional neural networks and the details of their hyperparameters and their connectivity.

• Convolution layer

The convolution layer is the core building block of a Convolutional Network that does most of the computational heavy lifting. The name of convolutional neural network also comes from this layer. Let's first discuss what the convolution layer computes without consideration of brain and neuron analogies. In the context of convolutional neural network, a convolution is a linear operation that involves the multiplication of a set of weights with the input, similar to a traditional neural network. Recall that the convolution layer was designed for two dimensional input, the multiplication is performed between an array of input data and a two dimensional array of weights which is called a filter or a kernel. The convolution layer consists of a set of trainable parameters. These trainable parameters are the weights of each filter. Every filter is small spatially along width and height, but extends through the full depth of the input volume. For example, a typical filter on a first layer of a convolutional neural network can have the size $5 \times 5 \times 3$. This size can be interpreted so that there are 5 pixels for width and height, and 3 because images have depth 3 which are the color channels. During the forward pass, this filter slides each filter across the width and height of the input volume and computes dot products between the entries of the filter and the input at any position. This process also calls convolve. In other words, this filter convolves across the width and height of the input volume. When the filter slides over the width and height of the input volume, a 2 dimensional activation map that gives the responses of that filter at every spatial position will be produced. Intuitively, the network will train filters that activate when they see some type of visual feature such as an edge of some orientation or a blotch of some color on the first layer, or eventually entire honeycomb or wheel-like patterns on higher layers of the network. Now, there is an entire set of filters in each convolution layer, and each of them will produce a separate 2 dimensional activation map. In our previous example, there are 12 filters in the convolution layer. These activation maps will be stacked along the depth dimension and produce the output volume.

The idea of applying the convolutional operation to image data is not new or unique to convolutional neural networks. It is a common technique used in computer vision to extract features like edges. Historically, computer vision researchers were designed filters by hand, and then applied to an image to result in a feature map or output from applying the filter then makes the analysis of the image easier in some way. Example of these designed filters are histograms of oriented gradients (HOG) in which some filters were designed to extract edges in specific orientations. For example, in the following there is a hand crafted 3×3 element filter for detecting vertical lines.

$$filter = \begin{bmatrix} 0 & 1 & 0 \\ 0 & 1 & 0 \\ 0 & 1 & 0 \end{bmatrix} \tag{1}$$

Applying this filter to an image will result in a feature map that only contains vertical lines. It is a vertical line detector. You can see this from the weight values in the filter which any pixel values in the center vertical line will be positively activated and any on either side will be negatively activated or considered as zero. Sliding this filter systematically across pixel values in an image can only highlight vertical line pixels. A horizontal line detector is the transposed of the above filter. Combining the results from both feature maps will result in all of the lines in an image being highlighted. A suite of tens or even hundreds of other small filters can be designed to detect other features in the image. The innovation of using the convolution operation in a neural network is that the values of the filter are weights to be trained during the training of the network. Therefore, the network will learn what types of features to extract from the input. Specifically, the network is forced to learn to extract features from the image that minimize the loss for the specific task the network is being trained to solve, for example extracting features that are the most useful for classifying images as dogs or cats. Sometimes some hand craft filters like HOG cannot be helpful to classify for example CIFAR-10 images. Therefore, you should design new filters and test whether there are good enough for your CIFAR-10 classification task. This is a tedious task which convolutional neural networks relieve us from it. Although learning a single filter specific to a machine learning task is a powerful technique, in convolutional neural networks there are multiple filters for different channels which make it more powerful. The first point about the convolutional neural networks is that they do not train a single filter. In contrast they train multiple filters which can extract multiple features in parallel for a given input like image. For example, it is common for a convolutional layer to learn from 32 to 512 filters in parallel for a given input. This gives the model 32, or even 512, different ways of extracting features from an input. In other words, this gives many different ways of both learning to see during the training phase, and many different ways of seeing the input data in testing phase. This diversity allows specialization, for example not just vertical and horizontal lines, but the specific lines seen in your specific training data. However, color images have multiple channels which is typically one for each color channel that are red, green, and blue. This means that a single image is input to the model which is a volume with depth 3. A filter must always have the same number of channels as the input. This is often referred to as depth. Therefore, if an input image has 3 channels or depth of 3, then a filter applied to that image must also have 3 channels or depth of 3. In this case, a 3 × 3 filter in fact is 3 × 3 × 3 for width, height, and depth. Regardless of the depth of the input and depth of the filter, the filter is applied to the input using a dot product operation which results in a single value. This means that if a convolutional layer has for example 32 filters, these 32 filters are not just two dimensional for the two dimensional image input, but also are three dimensional, having specific filter weights for each of the three channels. It is important to notice that each filter results in a single feature map. In other words, the depth of the output of applying the convolutional layer with 32 filters is 32 for the 32 feature maps created. In this way, the convolutional neural networks can learn the structure of data from each channel and combine them. Convolution layers are not only applied to input data, like raw pixel values, but they can also be applied to the output of other layers. The stacking of convolutional layers allows a hierarchical decomposition of the input. Consider that the filters that operate directly on the raw pixel values will learn to extract low level features, such as lines in different orientations. The filters that operate on the output of the first line layers may extract features that are combinations of lower level features, such as features that comprise multiple lines to express shapes like face, car, animal, or house. This process continues until very deep layers which combines these features and work on extracting faces, animals, houses features. The last layer of features is also called high level features. This is exactly what happens in practice. The abstraction of features to high and higher orders as the depth of the network is increased.

Like the neural network, the convolutional neural networks have interpretation in biology. Every entry in the 3D output volume can also be interpreted as an output of a neuron that looks at only a small region in the input and shares parameters with all neurons to the left and right spatially since these numbers all result from applying the same filter. As you saw previously, it is impractical to connect neurons to all neurons in the previous volume when dealing with high dimensional inputs such as images. Instead, each neuron connects to only a local region of the its input volume. The spatial extent of this connectivity is a hyperparameter called the receptive field of the neuron. This can also be interpreted as the size of each filter. The extent of the connectivity along the depth axis is always equal to the depth of the input volume. It is important to emphasize again that this asymmetry in how the spatial dimensions (which are width and height) and the depth dimension are treated. The connections are local in 2 dimensional space along with width and height, but always full along the entire depth of the input volume. Consider the CIFAR-10 image classification task. Suppose that the input volume has a size of 32 × 32 × 3. If the receptive field or the filter size is 5 × 5, then each neuron in the convolution layer will have weights to a $5 \times 5 \times 3$ region in the input volume, for a total of $5 \times 5 \times 3$ which is 75 weights and also one bias term. Note that the extent of the connectivity along the depth axis must be 3, since this is the depth of the input volume. For another example, consider the output volume size of a layer in convolutional neural network is $16 \times 16 \times 20$. Therefore, using an example receptive field size of 3×3, every neuron in the convolution layer have a total of $3 \times 3 \times 20$ which is 180 connections to the input volume plus one bias term. Note that, again, the connectivity is local in 2D space, but full along the input depth.

Until now, the local connectivity of each neuron in the convolution layer to the input volume in convolutional neural network is explained. There is also another question which is how many neurons are in the output volume or how they are arranged. To answer this question, you should consider three hyperparameters the depth, stride, and zero padding. These parameters control the size of the output volume. The depth of the output volume is a hyperparameter. It corresponds to the number of filters you would like to use, each learning to look for something different in the input. For example, if the first convolutional layer takes as input the raw image, then different neurons along the depth dimension may activate in presence of various oriented edges, or blobs of color. The second hyperparameter is the stride which slides the filter over the input volume. When the stride is 1 then the filters are moved one pixel at a time. When the stride is 2, then the filters jump 2 pixels at a time as the filters are slide around. This will produce smaller output volumes spatially. The third hyperparameter is zero padding. Sometimes it is convenient to pad the input volume with zeros around the border. Therefore, the size of this zero padding is a hyperparameter. The nice feature of zero padding is that it will allow us to control the spatial size of the output volumes. Now you are able to compute the spatial size of the output volume as a function of the input volume size (W), the receptive field size of the convolution layer neurons (F), the stride with which they are applied (S), and the amount of zero padding used (P) on the border. It is easy to show that the correct formula for calculating how many neurons fit is given by the following equation

$$\frac{W - F + 2P}{S} + 1 \qquad (2)$$

The reason to multiply the amount of zero padding is that both sides of the input volume are zero pad in each spatial direction. For example, if the input volume has size 7×7 and the convolution layer has a 3×3 filter with stride 1 and no zero padding (pad 0), the output volume has size of 5×5. It is easy to show that with stride 2 the output volume size is 3×3. In general, setting zero padding to be $P = \dfrac{F-1}{2}$ when the stride is S is equal to 1, ensures that the input volume and output volume will have the same size spatially. It is very common to use zero padding in this way. Note that the spatial arrangement hyperparameters have mutual constraints. For example, consider the situation in which the input has size W is 10, there is not using zero padding which means that P is 0, and the filter size F is equal to 3. Using these settings, it is impossible to use stride S equal to 2, since from the above equation the output volume spatial size is $\dfrac{10-3+0}{2}+1$ which is 4.5. This value is not an integer and indicating that the neurons do not fit neatly and symmetrically across the input. Therefore, this setting of the hyperparameters is considered to be invalid. When you build a convolutional neural network using a library like Tensorflow or PyTorch, they can throw an exception, or zero pad the rest to make it fit, or crop the input to make it fit. When you design the architecture of a convolutional neural network, sizing the layers like pooling and convolution appropriately so that all the dimensions work out can be a real headache. Here, with using the zero padding technique and some design guidelines will significantly alleviate the problem. One of the famous convolutional neural network is AlexNet which designed by Krizhevsky. This convolutional neural network won the ImageNet challenge in 2012. The input or input image size for this network is $227 \times 227 \times 3$. On the first convolution layer, it used neurons with receptive field size (or equivalently the filter size) F equal to 11, stride S is equal to 4 and no zero padding which means the P value is equal to 0. Also, this layer has a depth of K equal to 96. Therefore, the output volume of this convolution layer has size of $55 \times 55 \times 96$. Each of the $55 \times 55 \times 96$ neurons in this volume is connected to a region of size $11 \times 11 \times 3$ in the input volume. In addition, all 96 neurons in each depth column are connected to the same $11 \times 11 \times 3$ region of the input, but of course with different weights. In the actual paper, it claims that the input images are 224×224 which is obviously incorrect because $\dfrac{224-11}{4}+1$ that equal to 54.25 which is clearly not an integer.

The parameter sharing scheme is used in convolution layers to control the number of parameters. Using the real world example above, AlexNet, you see that there are $55 \times 55 \times 96$ which equal to 290,400 neurons in the first convolution layer, and each of them has $11 \times 11 \times 3$ which equal to 363 weights and 1 bias term. Together, this results to 290400×364 which is 105,705,600 parameters on only the first layer of the convolutional neural network. There are very high trainable parameters for the first layer of this network. If you consider one reasonable assumption, the number of trainable parameter can be reduced dramatically. This assumption is that if one feature is useful to compute at some spatial position $(x.y)$, then it should also be useful to compute at a different position (x_2, y_2). In other words, denoting a single two dimensional slice of depth as a depth slice (for example a volume of size $55 \times 55 \times 96$ has 96 depth slices which each of them has size 55×55), one can constrain the neurons in each depth slice to use the same weights and bias. With this parameter sharing scheme, the first convolution layer of this example now have only 96 unique set of weights which is one for each depth slice, and therefore in total, there is of $96 \times 11 \times 11 \times 3$ which is 34,848 unique weights and 96 bias terms, and finally 34,944 parameters. Alternatively, all 55×55 neurons in each depth slice will now be using the same parameters. In practice during backpropagation, every neuron in the volume will compute the gradient for its weights, but these gradients will be added up across each depth slice and only update a single set of weights per slice. Notice that if all neurons in a single depth slice are using the same weight vector, then the forward pass of the convolution layer can in each depth slice be computed as a convolution of the neuron's weights with the input volume. This the reason to name this layer as convolution layer. This is why it is common to refer to the sets of weights as a filter (or a kernel), that is convolved with the input. Notice that the parameter sharing assumption is reasonable. If detecting a horizontal edge is important at some location in the image, it should intuitively be useful at some other location as well due to the translationally invariant structure of images. Therefore, there is no need for it to relearn how to detect a horizontal edge at every one of the 55×55 distinct locations in the convolution layer output volume. Note that sometimes the parameter sharing assumption may not make sense. This is especially the case when the input images to a convolutional neural network have some specific centered structure, where you should expect that completely different features should be learned on one side of the image than another. One practical example is when the inputs are faces that have been centered in the image. You might expect that different eye specific or hair specific features could be learned in different spatial locations. In that case, it is common to relax the parameter sharing scheme and instead simply call the layer a locally connected layer.

Note that the convolution operation essentially performs dot products between the filters and local regions of the input. A common implementation pattern of the convolution layer is to take advantage of this fact and formulate the forward pass of a convolutional layer as one big matrix multiply. It is wise to use the efficient matrix libraries instead of writing for loops by yourself. Also, almost all of deep learning libraries in python leverage the matrix multiplication implementation on GPU and speed up the training and testing phases multiple times.

The other issue you must notice is the backward pass for backpropagation in training phase. The backward pass for a convolution operation for both the data and the weights is also a convolution. It is easy to derive in the

1 dimensional case as an example and extend it to 2 dimensional and higher dimensions.

You may at first be confused to see 1×1 convolutions especially when you come from a signal processing background. Normally signals are 2 dimensional so the 1×1 convolutions do not make sense. However, in convolutional neural networks, this is not the case because one must remember that the operation is done over 3 dimensional volumes, and that the filters always extend through the full depth of the input volume. For example, if the input volume size is $32 \times 32 \times 3$ then doing 1×1 convolutions would effectively be doing 3 dimensional dot products, since the input depth is 3 channels.

A recent development in convolutional neural networks is to introduce one more hyperparameter to the convolution layer, which is called the dilation. Until now, you consider that the filters in the convolution layer are contiguous. However, it is possible to have filters that have spaces between each cell, called dilation. This can be very useful in some settings to use in conjunction with zero dilated filters because it allows you to merge spatial information across the inputs much more aggressively than with fewer layers. For example, if you stack two 3×3 convolution layers on top of each other then, you can convince yourself that the neurons on the 2nd layer are a function of a 5×5 patch of the input. If you use dilated convolutions, then this effective receptive field would grow much quicker.

Let's summarize what you learn about the convolution layer until now. Consider that the input volume size is $W_1 \times H_1 \times D_1$. To specify the convolution layer, you should set the number of filters (K), their spatial extent (F), the stride (S), and amount of zero padding (P). By these settings, the output volume size is equal to $W_2 \times H_2 \times D_2$ where

$$W_2 = \frac{W_1 - F + 2P}{S} + 1 \tag{3}$$

$$H_2 = \frac{H_1 - F + 2P}{S} + 1 \tag{4}$$

$$D_2 = K \tag{5}$$

If using the parameter sharing assumption, the number of weights for each filter is equal to $F \times F \times D_1$, and in total there is $F \times F \times D_1 \times K$ weights and K biases. In the output volume, the d-th depth slice (which has size $W_2 \times H_2$) is the result of performing a valid convolution of the d-th filter over the input volume with a stride of S, and then offset by d-th bias. A common setting of the hyperparameters is F sets to 3, S sets to 1, and P sets to 1. However, there are common conventions and rules of thumb that motivate these hyperparameters. In general, it is best to set the stride S to 1. This is because the smaller strides work better in practice. In addition, if the stride is S to 1 you can leave all spatial down sampling to the pooling layers, while the convolution layer is only transforming the input volume depth wise.

By now, you what the convolution layer is and how it produces the output. When you design a convolutional neural network, you should set the number of filters and filter size for each layer. The question here is how to find the best value for these hyperparameters. One simple solution for this question is trial and error. You can set these hyperparameters and test which one has the best result. However, this is a tedious task. There is some intuition regarding the convolution layer that can help you to narrow your search grid. One of the important intuitive facts about the convolution layer is that smaller filters can collect and can extract local information while the bigger filters represent a more global, high level and representative information. For example, a filter with size 3 × 3 extracts the local features like edges and lines in different orientations while a filter with size 9 × 9 extracts more high level and global features. Remember that the first layers in a convolutional neural network extract local features while the deeper convolution layer extracts the more high level features. You can conclude that the first filter size of the first layers should be small like 3 × 3 to extract low level features while the filter size in the deeper layer should be large like 9 × 9 to combines these low level features and create a high level feature like a wheel, circle, and faces. There are not many low level features to extract, therefore if you can set the number of filters in the first layers small. However, the deeper layers have more options to extract high level features compared to the first layers. Hence, it is reasonable to consider more filter as the layer will be deeper.

The last issue about the convolution layer is how to pad the input layer. There are many techniques that can be used for padding such as valid padding, same padding, causal padding, constant padding, reflection padding, and replication padding. Most of these padding types are implemented in the deep learning process. In the valid padding, there is no padding. The other name of the valid padding is no padding. In this padding, the size of input image will be same. Another zero padding type is same padding. It is also known as zero padding. This padding ensures that the output has the same shape as the input data. Another type of padding that resembles same padding is constant padding. The outcome of constant padding can be the same which is what same padding does. Moreover, the shape of output is the same as the input. However, rather than pad with zero value, the constant padding allows you to pad with a desired constant value. Another type of padding is reflection padding. In the reflection padding, the values are padded with the reflection or mirror of the values directly in the opposite direction of the. Reflective padding sometime can improve the empirical performance of the model. Replication padding is very similar to reflection padding, but is slightly.

In the replication padding, instead of reflecting like a mirror, you simply take a copy, and mirror it.

• Pooling layer

It is common to periodically insert a pooling layer between successive convolution layers in a convolutional neural network architecture. Its function is to progressively reduce the spatial size of the representation to reduce the amount of parameters and computation in the network. One of the main goals in using the pooling layer is to reduce the feature maps and therefore the chance of overfitting. In other words, you can control overfitting by using the pooling layer. The pooling layer operates independently on every depth slice of the input and resizes it spatially.

Based on what we discussed about the convolutional layers, they prove very effective. When you stack convolutional layers in deep models, you allow the layers close to the input to learn low level features like lines or edges in different orientations and layers deeper in the model learn high level or more abstract features, like shapes or specific objects by combining the low level features which are extracted in the first layers. A limitation of the feature map output of convolutional layers is that they record the precise position of features in the input. This means that small movements in the position of the feature in the input image will result in a different feature map. This can happen with re cropping, rotation, shifting, and other minor changes to the input image. A common approach to addressing this problem from signal processing field is called down sampling. This is where a lower resolution version of an input signal is created that still contains the large or important structural elements, without the fine detail that may not be as useful to the task. As mentioned in convolution layer section, down sampling can be achieved with convolutional layers by changing the stride of the convolution across the image or the input volume of each layer. A more robust and common approach is to use a pooling layer. It is logical to put a pooling layer after the convolution. This is due to the fact that this layer reduces the dimension of convolution layer for the next layer by keeping the most important information. Remember that each neuron in convolution neural network should activate. The question may rise is that it is better to put the pooling layer before the activation layer or after this layer? It is shown that the best place for the pooling layer is after a nonlinearity function like ReLU has been applied to the feature maps output by a convolutional layer. If you check our first convolutional neural network example, you can see that the pooling layer comes after the activation layer. The addition of a pooling layer after the convolutional layer is a common pattern used for ordering layers within a convolutional neural network that may be repeated one or more times in a given model. Notice that the pooling layer operates upon each feature map separately to create a new set of the same number of pooled feature maps. Pooling involves selecting a pooling operation, much like a filter to be applied to feature maps. The size of the pooling operation or filter is smaller than the size of the feature map. Specifically, it is almost always 2 × 2 pixels applied with a stride of 2 pixels. This means that the pooling layer will always reduce the size of each feature map by a factor of 2, or in the other words, each dimension is halved, reducing the number of pixels or values in each feature map to one quarter the size. For example, a pooling layer applied to a feature map of 6 × 6 will result in an output pooled feature map of 3 × 3. You might understand that the pooling layer has two important hyperparameters which should set correctly. In addition, the pooling layer does not have any trainable parameters and only has some parameters which should be specified, rather than learned. There are two common functions used in the pooling operation which are, average operator and maximum operator. Average pooling calculates the average value for each patch on the feature map. For example, when you applied a 2 × 2 average pooling filter to a feature map with stride 1, it slides over the feature map and average elements in 2 × 2 area. In other words, average pooling involves calculating the average for each patch of the feature map. This means that each 2 × 2 square of the feature map is down sampled to the average value in the square. You can also assume that this layer uses convolution layer with known weights. In the average pooling with size 2 × 2, all of the weights are equal to $\frac{1}{4}$. The other type of pooling is maximum pooling (or Max pooling). The Max pooling calculate the maximum value for each patch of the feature map. Like the average pooling, you can consider this type of pooling as a convolution layer with known weights. Therefore, there isn't any trainable parameter in pooling layer. In other words, in Max pooling calculates the maximum value in each patch of each feature map. The results are down sampled or pooled feature maps that highlight the most present feature in the patch, not the average presence of the feature in the case of average pooling. This has been found to work better in practice than average pooling for computer vision tasks like image classification. The result of using a pooling layer and creating down sampled or pooled feature maps is a summarized version of the features detected in the input. They are useful as small changes in the location of the feature in the input detected by the convolutional layer will result in a pooled feature map with the feature in the same location. This capability added by pooling is called the invariance of model to local translation. Invariance to translation means that if we translate the input by a small amount, the values of most of the pooled outputs do not change. There is another type of pooling that is sometimes used called global pooling. Instead of down sampling patches of the input feature map, global pooling down samples the entire feature map to a single value. This would be the same as setting the pooling size to the size of the input feature map. Global pooling can be used in a model to aggressively summarize the presence of a feature in an image. It is also sometimes used in models as an alternative to using a fully connected layer to transition from feature maps to an output prediction for the model.

In general, the most common type of pooling layer is Max pooling and the most common pooling size is filters with size of 2×2 applied with a stride of 2 down samples every depth slice in the input by 2 along both width and height. Every maximum operation would in this case be taking a maximum over 4 numbers in the 2×2 region in some depth slice. The depth dimension remains unchanged. In nutshell, consider the input volume of the pooling layer is equal to $W_1 \times H_1 \times D_1$. There are two hyperparameters that should be set to specify the pooling layer. These two hyperparameters are the spatial extent F, and the stride S. The output volume size of the pooling layer is equal $W_2 \times H_2 \times D_2$ to where:

$$W_2 = \frac{W_1 - F}{S} + 1 \tag{6}$$

$$H_2 = \frac{H_1 - F}{S} + 1 \tag{7}$$

$$D_2 = D_1 \tag{8}$$

Again note that the pooling layers have zero trainable parameters since it uses a fixed function. In other words, the weights of filters used in pooling layer is known and they do not need to train. Also, it is not common to pad the input using zero padding in pooling layers. It is worth it to note that there are only two commonly seen variations of the max pooling layer found in practice. The first commonly used pooling layer has spatial extent F equal to 3, and the stride S equal to 2. This pooling layer has overlapping and therefore sometimes called overlapping pooling. The second commonly used pooling layer which is more commonly has spatial extent F equal to 2, and the stride S equal to 2. Note that this pooling layer discards exactly 75% of the activations in an input volume. Remember that the pooling sizes with larger receptive fields can be too destructive.

In addition to max pooling, the pooling units can also perform other functions, such as average pooling or even L2-norm pooling. Average pooling was often used historically but has recently fallen out of favor compared to the max pooling operation, which has been shown to work better in practice. Recall from the backpropagation section that the backward pass for a maximum function like $max(x.y)$ has a simple interpretation as only routing the gradient to the input that had the highest value in the forward pass. Therefore, during the forward pass of a pooling layer it is common to keep track of the index of the max activation so that gradient routing is efficient during backpropagation.

Many researchers do not like the pooling operation and think about how they can get away without it. One solution is striving for simplicity. All the convolutional neural network propose to discard the pooling layer in favor of architecture that only consists of repeated convolution layers. To reduce the size of the representation they suggest using a larger stride in the convolution layer once in a while. Discarding pooling layers has also been found to be important in training good generative models, such as variational autoencoders (VAEs) or generative adversarial networks (GANs). It seems likely that future architecture will not use the pooling layers.

• Activation layer

The activation layer or nonlinearity layer in a convolutional neural network consists of an activation function that takes the feature map generated by the convolution layer and creates the activation map as its output. The activation function is an element wise operation over the input volume and therefore the dimensions of the input and the output are identical. Similar to neural network which discussed in the previous chapters, the activation function can be ReLU, leaky ReLU, sigmoid, Tanh, or the softmax function. However, like the neural network, more recent research suggests the ReLU activation layer is advantageous over the sigmoid and Tanh activation layer in convolutional neural networks. Also, remember that the softmax activation function is usually used in the last fully connected layer because the output of the activation function is a probability value. Please refer to the neural network section for more details.

• Fully connected layer

The name of the fully connected layer aptly describes itself. If you remember, the pixel values of the input image are not directly connected to the output layer in partially connected layers in the convolution layer. However, in the fully connected layer, each node in the output layer connects directly to a node in the previous layer. This layer performs the task of classification based on the features extracted through the previous convolution layers with their different filters. Note that the convolutional and pooling layers tend to use the ReLU activation function, while the fully connected layers usually leverage a softmax activation function to classify inputs appropriately. The output of the last layer is that the fully connected layer produces a probability from 0 to 1 for each class. For example, in CIFAR-10 dataset the number of classes is 10 and therefore the last layer of fully connected layer has 10 neurons which each represents the probability of being of that class.

The convolution and fully connected layers are very similar to each other. One of the differences between fully connected and convolution layers is that the neurons in the convolution layer are connected only to a local region in the input volume. The other difference is that the neurons in convolution volume share parameters. However, the neurons in both layers still compute dot products and you can conclude that their functional form is identical. Therefore, it is possible to convert between fully connected and convolution layers. For any convolution layer there is a fully connected layer that implements the same forward function. The weight matrix would be a large matrix that is mostly zero except for at certain blocks where the weights in many of the blocks are equal (due to parameter sharing). The reason to most elements of the matrix should be zero is the local connectivity. When the additional matrix elements are set to zero, then the you cut of the connectivity of neuron to all of the input volume. Also, the reason of equality of weights in many of blocks is the parameter sharing assumption. Conversely, any fully connected layer can be converted to a convolution layer. For example, a fully connected layer with K equal to 4096 that is looking at some input volume of size $7 \times 7 \times 512$ can be equivalently expressed as a convolution layer with size of filter F equal to 7, zero padding P equal to 0, stride S equal to 1, and number of filters K equal to 4096. In other words, the filter size is set to be exactly the size of the input volume, and hence the output will simply be $1 \times 1 \times 4096$ since only a single depth column fits across the input volume. The result of this convolution layer is identical with the initial fully connected layer.

From these two conversions, the ability to convert a fully connected layer to a convolution layer is particularly useful in practice. Consider a convolutional neural network architecture that takes a $224 \times 224 \times 3$ image, and uses a series of convolution layers and pooling layers to reduce the image to an activations volume of size $7 \times 7 \times 512$. From there, the AlexNet network uses two fully connected layers of size 4096 and finally the last fully connected layers with 1000 neurons that compute the class scores. You can convert each of these three fully connected layers to convolution layers. In the first fully connected layer, you can replace with a convolution layer that uses filter size F equal to 7, giving output volume $1 \times 1 \times 4096$. For the second fully connected layer, you can replace it with a convolution layer that uses filter size F equal to 1, giving output volume $1 \times 1 \times 4096$. Similarly, you can replace the last fully connected layer with filter size F equal to 1, and giving the final output $1 \times 1 \times 1000$. Each of these conversions could in practice involve manipulating (which is reshaping) the weight matrix W in each fully connected layer into convolution layer filters. It turns out that this conversion allows the network to slide the original convolutional neural network very efficiently across many spatial positions in a larger image, in a single forward pass. For example, if image with size of 224 × 224 gives a volume of size 7 × 7 × 512, then forwarding an image of size 384 × 384 through the converted architecture would give the equivalent volume in size 12 × 12 × 512. Following through with the next 3 convolution layers that we just converted from fully connected layers now gives the final volume of size 6 × 6 × 1000. Note that instead of a single vector of class scores of size 1 × 1 × 1000, there is an entire 6 × 6 array of class scores across the 384 × 384 image. Evaluating the original convolutional neural network with FC layers, independently across 224 × 224 crops of the 384 × 384 image in strides of 32 pixels gives an identical result to forwarding the converted convolutional neural network one time. Naturally, forwarding the converted convolutional neural network a single time is much more efficient than iterating the original convolutional neural network over all those 36 locations, since the 36 evaluations share computation. This trick is often used in practice to get a better performance, where for example, it is common to resize an image to make it bigger, use a converted convolutional neural network to evaluate the class scores at many spatial positions and then average the class scores. The last note is that what if you wanted to efficiently apply the original convolutional neural network over the image but at a stride smaller than 32 pixels? you can achieve this with multiple forward passes. For example, note that if you wanted to use a stride of 16 pixels you can do so by combining the volumes received by forwarding the converted convolutional neural network twice. The first over the original image and the second over the image but with the image shifted spatially by 16 pixels along both width and height.

• Dropout layer

One of the other layers in a convolutional neural network is dropout layer. As you remember, dropout is a technique used to prevent a model from overfitting. Dropout works by randomly setting the outgoing edges of hidden neurons to 0 at each update of the training phase. In convolutional neural networks, you can use the dropout for the fully connected layers. For the convolutional layers, they usually use the pooling layer to reduce the number of trainable parameters and reduce the chance of overfitting. Please refer to the neural network section for more details.

• Batch normalization layer

Training deep neural networks with tens of convolution and fully connected layers is challenging. As mentioned in the neural network chapter, the training process can be sensitive to the initial random weights and configuration of the learning algorithm. One possible reason for difficulty in training phase is the distribution of the inputs to layers deep in the network may change after each mini batch when the weights are updated. This can cause the learning algorithm to forever chase a moving target. This change in the distribution of inputs to layers in the network is referred to by the technical name internal covariate shift. Batch normalization is a technique for training very deep neural networks that standardizes the inputs to a layer for each mini batch. This has the effect of stabilizing the learning process and dramatically reducing the number of training epochs that are required to train deep networks. Batch normalization accelerates training, in some cases by halving the epochs or better, and provides some regularization, reducing generalization error.

One aspect of challenges in training a convolutional neural networks, is that the model is updated layer by layer backward from the output to the input using an estimate of error. There is one assumption when the weights update and that is assumes the weights in the layers prior to the current layer are fixed. In other words, using the gradient you can update the parameters but under the assumption that the other layers do not change, while in practice the updating process is done simultaneously for all the layers of the convolutional neural network. Because all layers are changed during an update, the update procedure is forever chasing a moving target. For example, the weights of a layer are updated and given an expectation that the prior layer outputs values with a given distribution. This distribution is likely changed after the weights of the prior layer are updated. In other words, the distribution of each input layer changes in the training process when the parameters of the previous layers change. To combat this problem, you should initialize the learning rate with a low value and initialize the weights and biases carefully. This causes the training process slow down and also make it hard to train models due to the saturating in nonlinear activation functions like sigmoid and Tanh. Batch normalization is proposed as a technique to help coordinate the update of multiple layers in the model. It does this by scaling the output of the layer, specifically by standardizing the activations of each input variable per mini batch, such as the activations of a node from the previous layer. Recall that standardization refers to rescaling the data to have a mean of zero and a standard deviation of one. Sometimes, this process is called whitening. Using the standardization, to help the input volume for each layer in convolutional neural network has same distribution and therefore, the main assumption in the gradient descent will hold. In other words, with standardizing the activations of the prior layer, the assumptions of the subsequent layer should have a close distribution of inputs during the weight update which will hold. This has the effect of stabilizing and speeding up the training process of deep convolutional neural networks. In a nutshell, the batch normalization standardizes only the mean and variance of each neuron in order to stabilize training process, but it allows the relationships between neurons and the nonlinear activation function of a single unit to change. Normalizing the inputs of each layer can dramatically reduce the number of epochs required in the training process. It can also have a regularizing effect, reducing generalization error much like the use of activation regularization. This layer has a dramatic effect on solved optimization especially for convolutional neural networks with nonlinear activation functions like sigmoid and Tanh. Although reducing internal covariate shift was a motivation in the development of the method, there is some suggestion that instead batch normalization is effective because it smooths and, simplifies the optimization function that is being solved when training the network.

Batch normalization can be implemented during training by calculating the mean and standard deviation of each input variable to a layer per mini batch and using these statistics to perform the standardization. Alternately, a running average of mean and standard deviation can be maintained across mini batches, but may result in unstable training. After training, the mean and standard deviation of inputs for the layer can be set as mean values observed over the training dataset. For small mini batch sizes or mini batches that do not contain a representative distribution of examples from the training dataset, the differences in the standardized inputs between training and test can result in noticeable differences in performance. This can be addressed with a modification of the method called batch renormalization that makes the estimates of the variable mean and standard deviation more stable across mini batches. In other words, the batch renormalization extends the batch normalization to ensure that the activations match between training and test phases. During the training phase, the batch normalization layer does the following steps. In the first step, the mean and variance of the input volume of this layer are calculated by the following equations.

$$\mu_B = \frac{1}{m} \sum_{i=1}^{m} x_i \tag{9}$$

$$\sigma_B^2 = \frac{1}{m} \sum_{i=1}^{m} (x_i - \mu_B)^2 \tag{10}$$

The first equation is referred to as batch mean while the second equation is referred to as batch variance. In the above equation, the m is the number of samples. In the second step, the input volume is normalized using the mean and variance calculated from the previous step. To normalize the input volume, you can use the following equation.

$$\bar{x}_i = \frac{x_i - \mu_B}{\sqrt{\sigma_B^2 + \epsilon}} \tag{11}$$

The ϵ in the above equation is a very small number which avoid to zero dividing. In the last step, the $\bar{x}_i$ is scaled and shifted in order to obtain the output of the layer as the following equation.

$$y_i = \gamma \bar{x}_i + \beta \tag{12}$$

where γ and β in the above equation is the scale and shift factor, respectively. Notice that γ and β are trainable parameters which learned during the training process along with the original parameters of the network. Therefore, the batch normalization layer has two trainable parameters. Do not forget that the statistics of the input volume (mean and variance) during the test phase are fixed. They are estimated using the previously calculated means and variances of each training batch.

The standardization of inputs can be applied to input variables for the first hidden layer or to the activations from a hidden layer for deeper layers. In practice, it is common to allow the layer to learn two new parameters (γ and β) that allow the automatic scaling and shifting of the standardized layer inputs. It is due to the fact that simply normalization of the input volume can change what the layer can represent. Therefore, these parameters train in the training phase to restore the representation power of the convolutional neural network. Importantly the backpropagation algorithm is updated to operate upon the transformed inputs, and error is also used to update the new scale and the shifting parameters learned by the model. The standardization usually applies to the output volume of the convolution layer and before the activation layer. For example, the ResNet model which is one of the greatest convolutional neural networks, use the batch normalization layer after the convolutional layers in their very deep model.

When you are using the batch normalization in your convolutional neural network, you should consider some tips. Here, we provide some tips and suggestions for using the batch normalization layer. The first issue is that the batch normalization is a general technique that can be used to normalize the inputs to a layer. Therefore, you can use this layer in other types of networks like recurrent neural networks. The other tip is that where to use the batch normalization layer. The batch normalization layer can be used after or before the activation layer. It may be more suitable to use the batch normalization after the activation layer if the activation function has an S-shape format. For example, when you use the sigmoid or Tanh activation function, it is the best to use the batch normalization layer after the activation layer. But, for non-Gaussian distributions like the ReLU or leaky ReLU activation layer, it is the best to use the batch normalization before the activation layer. For example, in a paper in which was published in 2015, it is stated that they use the batch normalization before the activation layer (or nonlinearity) because matching the first and second moments is more likely to result in a stable distribution. You can test both cases and validate the model on your validation dataset to select the best solution. When you are using the batch normalization, the training procedure is more stable and faster. The faster training also means that the decay rate used for the learning rate may be increased. Therefore, you can choose the larger learning rate than when you do not use the batch normalization. This also speed up the optimization algorithm to find the best parameters. Remember that deep neural networks can be sensitive to the weights initialization. One of the advantages of using batch normalization in deep convolutional neural network is that the network is less sensitive to weight initialization. In other words, the stability to training brought by batch normalization can make training deep networks less sensitive to the choice of weight initialization method. The batch normalization layer can also be used to standardize raw input variables that have differing scales. If the mean and standard deviations calculated for each input feature are calculated over the mini batch instead of over the entire training dataset, then the batch size must be sufficiently representative of the range of each variable. Remember that the batch normalization layer has some regularization effect, reducing generalization error. Therefore, you can sometimes remove the dropout layer because the dropout layer has the regularization effect. There is an experiment that shows that if you remove the dropout layer from the Inception network and add the batch normalization layer, the overfitting does not increase while the training process speeds up. Further, it may not be a good idea to use batch normalization and dropout in the same network. The reason is that the statistics used to normalize the activations of the prior layer may become noisy given the random dropping out of nodes during the dropout procedure. But remember that there is not any clear answer to this question that can you use the batch normalization and dropout layers together in a network. You can test and try for your problem and discover which one is working better on your problem.

By now, you are familiar with the basic concept of convolutional neural network. To train a deep network you need to have a large dataset and appropriate resources like RAM and GPU. In practice, very few researchers train an entire convolutional neural network from scratch with random initialization, because it is relatively rare to have a dataset of sufficient size and suitable resources. Instead, it is common to pretrain a convolutional neural network on a very large dataset. For example, the ImageNet dataset in compute vision field, has 1.2 million images with 1000 categories. Afterwards, use this network either as an initialization or a fixed feature extractor for the task of interest. This idea is called transfer learning. There are three major transfer learning scenarios. The first scenario is use the convolutional neural network as a fixed feature extractor. Take a convolutional neural network which pretrained on for example ImageNet dataset, remove the last fully connected layer. In the original network the last layer has 1000 outputs which corresponds to the class scores. Afterwards, treat the rest of the network as a fixed feature extractor for the new dataset. You can extract features from this network and train a fully connected layer (or equivalently the neural network), support vector machine or any other classifier to classify the new images in your custom dataset. For example, in the AlexNet network, the feature extractor computes a 4096 dimensional vector for every image that contains the activations of the hidden layer immediately before the classifier. These features are called CNN codes. It is important for performance that these codes are ReLU if they were also part of the threshold during the training of the convolutional neural network on ImageNet dataset. Once you extract the 4096 dimensional codes or features from the AlexNet network for all images, you can train a linear classifier like linear SVM or softmax classifier for the new dataset. The second scenario is to fine tuning the convolutional neural network. The second strategy is to not only replace and retrain the classifier on top of the convolutional neural network on the new dataset, but to also fine tune the weights of the pretrained network by continuing the backpropagation. It is possible to fine tune all the layers of the original network, or it is possible to keep some of the earlier layers fixed or freeze them due to overfitting concerns, and only fine tune some higher level portion of the network. When your training dataset is not enough to train a large convolutional neural network, you can fine tune only some deep layers to avoid overfitting. You can also fine tune the first layers, but usually it is not best practice. Because the first layers extract low level features like edges and lines in different orientations. You can consider that most of the networks extract nearly the same low level features. The important issue is how to combine them to get the proper features for the task. Therefore, it is the best practice to train the last convolution layers in this situation. In other words, the first convolution layers contain more generic features that should be useful to many tasks, but deeper convolutional layers of in convolutional neural network becomes progressively more specific to the details of the classes contained in the original dataset. The third scenario is use the pretrained models. Since modern convolutional neural networks take a few weeks to train across multiple GPUs on ImageNet, it is common to release their final network checkpoints for the benefit of others who can use the networks for fine tuning. For example, the Caffe library has a Model Zoo where people share their network weights.

There are some questions that may rise. The first one is when and how to fine tune? The second one is how does someone decide what type of transfer learning should perform on a new dataset? The response to these questions depends on several factors, but the two most important ones are the size of the new dataset whether small or large, and its similarity to the original dataset. For example, the dataset is similar to ImageNet in terms of the content of images and the classes, or very different, such as microscope images. Again note that the features of convolutional neural networks are more generic in early layers and more original dataset specific in the deeper layers. Here are some common rules of thumb for navigating the four major scenarios. Consider that the new dataset is small and similar to original dataset. Since the dataset is small, it is not a good idea to fine tune the convolutional network due to the overfitting concerns. Since the data is similar to the original data, you can expect higher level features in the convolutional network to be relevant to this dataset as well. Therefore, the best idea might be to train a linear classifier on the CNN codes (or features). If the new dataset is large and similar to the original dataset the scenario is changed. Since you have more data in the dataset, you can have more confidence that the convolutional network does not overfit if you try to fine tune through the full network. In this situation, you can only fine tune the deep convolution layers and freeze the first layers. The third scenario is that the new dataset is small and very different from the original dataset. Since the data is small, it is likely best to only train a linear classifier. Since the dataset is very different, it may not be best to train the classifier from the top of the network, which contains more dataset specific features. Instead, it might work better to train the SVM classifier. The fourth scenario is that the new dataset is large and very different from the original dataset. Since the dataset is very large, you can expect that you can afford to train a convolutional neural network from scratch. However, in practice it is very often still beneficial to initialize with weights from a pretrained model. In this case, you have enough data and confidence to fine tune through the entire network.

There are a few additional tips when using the transfer learning technique. Note that if you want to use a pre-trained network, you might be slightly constrained in terms of the architecture you can use for your new dataset. For example, you cannot arbitrarily take out convolution layers from the pre-trained network. However, some changes are straight forward. For example, due to parameter sharing idea, you can easily run a pre-trained network on images of different spatial size. This is clearly evident in the case of convolution and pooling layers because their forward function is independent of the input volume spatial size. In case of fully connected layers, this still holds true because fully connected layers can be converted to a convolutional layer as we discussed before. For example, in the AlexNet network, the final pooling volume before the first fully connected layer is of size $6 \times 6 \times 512$. Therefore, the fully connected layer looking at this volume is equivalent to having a convolutional layer that has a receptive field size 6×6, and is applied with a padding of 0. The other issue is about the learning rates. It is common to use a smaller learning rate for convolutional neural network weights that are being fine-tuned, in comparison to when the weights are randomly initialized for the new linear classifier that computes the class scores of your new dataset. This is because you expect that the convolutional neural network weights are relatively good, so you do not wish to distort them too quickly and too much, especially while the new linear classifier above them is being trained from random initialization.

To understand and visualize the convolutional neural network layers, there are several approaches. In the following, each approach will be described. The first approach is to visualise the activations and first layer weights. This approach is one of the most straight forward visualization technique. Using this technique, the activations of the network during the forward pass is shown. For ReLU activation layers, the activations in the first epochs usually looks relatively blobby and dense, but as the training progresses the activations usually become sparser and localized. If you see that the output of some activation maps are zeros for many different inputs, you can guess that there is a bug in training process which cause the filter for these features map to be dead and can be a symptom of high learning rates. The other strategy is to visualize the weights of each layer. Note that these are usually most interpretable on the first convolution layer which is looking directly at the raw pixel data, but it is possible to also show the filter weights deeper in the network. The weights are useful to visualize because well trained networks usually display nice and smooth filters without any noisy patterns. Noisy patterns can be an indicator of a bug in the training phase. For example, the network does not been trained for long enough, or use a very low regularization strength that may have led to overfitting. Another visualization technique is to take a large dataset of images, feed them through the network and keep track of which images maximally activate some neuron. Therefore, you can visualize the images to get an intuition of what the neuron is looking for in its receptive field. One problem with this approach is that ReLU neurons do not necessarily have any semantic meaning by themselves. Rather, it is more appropriate to think of multiple ReLU neurons as the basis vectors of some space that represents in image patches. In other words, the visualization is showing the patches at the edge of the cloud of representations, along the axes that correspond to the filter weights. This can also be seen by the fact that neurons in a convolutional neural network operate linearly over the input space. Convolutional neural networks can be interpreted as gradually transforming the images into a representation in which the classes are separable by a linear classifier. You can get a rough idea about the topology of this space by embedding images into two dimensions so that their low dimensional representation has approximately equal distances rather than their high dimensional representation. There are many embedding methods that have been developed with the intuition of embedding high dimensional vectors in a low dimensional space while preserving the pairwise distances of the points. Among these, the t-SNE network is one of the best known methods that consistently produces visually pleasing results. To produce an embedding, you can take a set of images and use the convolutional neural network to extract the CNN codes (or CNN features). For example, the AlexNet network features are a 4096 dimensional vector. You can then plug these into the t-SNE network and get 2 dimensional vector for each image. The corresponding images can them be visualized in a grid. The other approach to get some intuition about your convolutional neural network is occluding different parts of the image. Suppose that a convolutional neural network classifies an image as a human. How can you be certain that it is actually picking up on the human in the image as opposed to some contextual cues from the background or some other miscellaneous object? One way of investigating which part of the image some classification prediction is coming from is by plotting the probability of the class of interest like human class as a function of the position of an occluder object. Using this procedure, you can iterate over regions of the image, set a patch of the image to be all zero, and look at the probability of the class. Therefore, you can visualize the probability as a 2 dimensional heat map.

By now, you are familiar with basics of convolutional neural networks. You also learn how to use the transfer learning to use a convolutional neural network for your own task. Now, it is the best time to introduce some of the popular and state of the art convolutional neural network. Please remember that in order to learn more about the convolutional neural networks, we highly recommend that you to read the original documentation of the convolutional neural networks. There are several architectures in the field of convolutional neural networks that are very powerful and famous. The most common networks are LeNet, AlexNet, GoogLeNet, VGGNet, and ResNet. The LeNet network is the first successful applications of convolutional neural networks which developed in 1990. This network has a simple architecture and is a good example to implement it with the famous deep learning libraries like Tensorflow and PyTorch. This network has a single convolution layer which immediately followed by a pooling layer. The other convolutional neural network which cause a great achievement in deep learning is the AlexNet. This is the first network that popularized convolutional neural networks in computer vision and was developed in 2012. The AlexNet was submitted to the ImageNet ILSVRC challenge and significantly outperformed the second competitor. The AlexNet network has a very similar architecture to LeNet network, but it is deeper, bigger, and convolution layers stacked on top of each other to extract more precise and robust features. The other network is ZFNet which is the improved version of the AlexNet. This network is an improvement on AlexNet by tweaking the architecture hyperparameters, in particular by expanding the size of the middle convolutional layers and making the stride and filter size on the first layer smaller. We discussed the filter size of first convolution layers in the convolution part. The next great network is GoogLeNet which introduced by Szegedy from Google. Its main contribution was the development of an Inception Module that dramatically reduced the number of parameters in the network. More precisely, in GoogLeNet the number of parameters is 4 million while in the AlexNet this number is 60 million. Additionally, this network uses average pooling instead of fully connected layers at the top of the convolutional neural network, eliminating a large amount of parameters that do not seem to be necessary for classification. There are also several versions of GoogLeNet is published which the most recent one is Inception-v4. These are two networks have great impact in compute vision classification problem. These networks are VGGNet and ResNet. The VGGNet is introduced in 2014 by Karen Simonyan and Andrew Zisserman. Its main contribution is in showing that the depth of the network is a critical component to achieve good performance. Their final best network contains 16 convolutions and fully connected layers and features an extremely homogeneous architecture that only performs 3×3 convolutions and 2×2 pooling from the beginning to the end. Their pretrained model is available in internet. You can use this network and train a new VGGNet for your problems with minimum resources. You can use the original network for feature extraction or fine tuning it for your task. One of the disadvantage of the VGGNet is that it is more expensive to evaluate and uses a lot more memory and parameters because it has 140 million trainable parameters. Most of these parameters are in the first fully connected layer, and it is since found that these fully connected layers can be removed with no performance downgrade, significantly reducing the number of necessary parameters. The other convolutional neural network is ResNet which it stands for Residual Network. The ResNet is developed by Kaiming in 2015. It features special skip connections and a heavy use of batch normalization. The architecture is also missing fully connected layers at the end of the network. ResNets are currently by far state of the art convolutional neural network models and are the default choice for using convolutional neural networks in practice. In particular, there are more recent developments that tweak the original architecture. Also, in many computer vision fields like face recognition or face matching, the backbone of the network is ResNet and only this network is fine tuned for the specific task [1, 12, 20].

Recurrent neural networks

Recurrent neural networks (RNNs) [77] consist of interconnected cycles that permit the outputs of the Long Short Term Memory Network (LSTM) to be presented to the input of the current state. It can also memorize the previous states such that the states passed earlier can be presented to the input of the new state. RNNs have many applications and can be used for time-series data, including, but not limited to, machine translation, candlestick prediction in stock trades, and speech recognition.

Recurrent neural networks are the state of the art algorithm for sequential data. These networks are used by many famous companies. For example, Apple's Siri and Google's voice search use these networks. Recurrent neural networks are another type of neural network. In the normal neural networks and convolutional neural networks, the input does not remember. While the recurrent neural networks are the first algorithms that remember their input, due to internal memory, this makes them perfectly suited for machine learning problems that involve sequential data. It is one of the algorithms behind the scenes of the amazing achievements seen in deep learning over the past few years.

The recurrent neural networks are a powerful and robust type of neural network and belong to the most promising algorithms in use because it is the only one with internal memory. Like many other deep learning algorithms, recurrent neural networks are relatively old. They were initially created in the 20 century, but thanks to the development of computational hardware like GPU, in recent years you can see their true potential. An increase in computational power along with the massive amounts of data, and the invention of long short-term memory networks (LSTM) in the last decade in the 20 century has brought recurrent neural networks to the foreground.

Because of their internal memory, recurrent neural networks can remember important things about the input they have received, which allows them to be very precise in predicting what is coming next. This is why they are the preferred algorithm for sequential data like time series, speech, text, financial data, audio, video, weather, and so on. Recurrent neural networks can form a much deeper understanding of a sequence and its context compared to other algorithms. As you remember, the convolutional neural networks are suitable for two-dimensional data like images while the recurrent neural networks are more suitable for three-dimensional data like time series and video. In other words, when the input data is a sequence of data that has temporal dynamics that connects the data sequence together and also there are some patterns in the temporal dynamics alongside the spatial pattern, you should use the recurrent neural networks. Since recurrent neural networks are used in the behind applications like Siri and Google Translate, the recurrent neural networks have a usage in the everyday life of most of most people.

To understand recurrent neural networks properly, you will need a working knowledge of normal feed forward neural networks and sequential data.

Sequential data is just ordered data in which related things follow each other. Examples of sequential data are financial data, the DNA sequence, or a video. The most popular type of sequential data is perhaps time series data, which is just a series of data points that are listed in time order. Recurrent neural networks and feed forward neural networks get their names from the way they channel information. In a feed forward neural network, the information only moves in one direction which is from the input layer to the output layer by passing the hidden layers. The information moves straight through the network and never touches a node twice. Feed forward neural networks or convolutional neural networks do not have the memory of the input they receive and are bad at predicting what is coming next. Because a feed forward network only considers the current input and does not have the notion of order in time. It simply cannot remember anything about what happened in the past except its training. In recurrent neural networks, the information cycles through a loop. When it makes a decision, it considers the current input and also what it has learned from the inputs it received previously.

A usual recurrent neural network has a short-term memory. Another good way to illustrate the concept of memory in a recurrent neural network is to explain it with an example. Consider you have a normal feed forward neural network and give it the word "neuron" as an input and it processes the word character by character. By the time it reaches the character "r," it has already forgotten about the previous letters which are "n," "e" and "u," and therefore this forgetting makes it almost impossible for this type of normal neural network to predict which character would come next. However, recurrent neural networks can remember those characters because of their internal memory. It produces output, copies that output, and loops it back into the network. In other words, recurrent neural networks can add the immediate past to the present. Therefore, recurrent neural networks have two inputs. The first one is the present and the second one is the recent past. This is important because the sequence of data contains crucial information about what is coming next, which is the reason for the power of the recurrent neural networks compared to the normal neural network and convolutional neural network. As you remember, the normal convolutional neural networks use 2D filters. Some researchers, use 3D filters in the convolutional neural network to enable these networks to work with sequential data. For example, some recent research uses the 3D convolutional neural networks for video understanding and action recognition. Despite these efforts, the recurrent neural networks are better than 3D convolutional neural networks.

Like neural networks and convolutional neural networks, a feed forward recurrent neural network assigns a weight matrix to its inputs and then produces the output. Note that recurrent neural networks apply weights to the current and also to the previous input. Furthermore, a recurrent neural network will also tweak the weights for both gradient descent and backpropagation over time. In addition, note that while feed forward neural networks map one input to one output, the recurrent neural networks can map one to one, one to many, many to many, and many to one. In one to one recurrent neural networks, the input and output are only one unit. While in the one to many recurrent neural networks the input is one and the output is many units, in many to one recurrent neural networks the input is many units and the output is one unit, and finally in the many to many recurrent neural networks in both of input and output are many units.

In neural networks, the forward propagation gets the output of the model and check whether the output is correct or incorrect. In contrast, the backpropagation is going backward through the neural network to find the partial derivatives of the error with respect to the weights. Finally, by using one of the optimization algorithms like mini batch gradient descent or Adam, the weights will update. Those derivatives are then used by gradient descent, an algorithm that can iteratively minimize a given function. Then it adjusts the weights up or down, depending on which it decreases the error. That is exactly how a neural network learns during the training process. Therefore, by using backpropagation the weights of the model are tweak during the training. The backpropagation through time is just an expression for backpropagation on for unrolled recurrent neural network. Unrolling is a visualization and conceptual tool, which helps us to understand what is going on within the recurrent neural network. Most of the time when implementing a recurrent neural network in the common programming frameworks, backpropagation is automatically taken care of, but there is a need to understand how it works to troubleshoot problems that may arise during the development process. You can consider recurrent neural networks as a sequence of neural networks that can be trained one after another with backpropagation. The conceptualization of unrolling is required for backpropagation through time since the error of a given time step depends on the previous time step. Within the backpropagation, through time the error is backpropagated from the last to the first time step while unrolling all the time steps. This allows calculating the error for each time step, which allows updating the weights. Note that backpropagation through time can be computationally expensive when you have a high number of time steps.

There are two major issues about the recurrent neural networks which should deal with them. Before explaining these issues, let's talk about the gradient in more detail. A gradient is a partial derivative with respect to its inputs. In other words, a gradient measures how much the output of a function changes if you change the inputs a little bit. Also, the gradient can be considered the slope of a function. The higher gradient value means the steeper the slope and the faster a model can learn. However, if the slope is zero, the model stops learning. A gradient simply measures the change in all weights with respect to the change in error. The first issue is gradient exploding. Exploding gradients are when the algorithm, without much reason, assigns stupidly high importance to the weights. Fortunately, this problem can be easily solved by truncating or squashing the gradients. The second issue is the gradient vanishing. Vanishing gradients occur when the values of a gradient are too small and the model stops learning or takes way too long as a result. Fortunately, it was solved through the concept of long short-term memory.

Long short term memory networks

Long short term memory networks (LSTMs) [76] are the special kinds of RNNs, which can learn the long and short dependencies of time-series data. Using past information is useful in many applications, depending on time, such as trading and economic data. They are suitable for time-series data due to remembering the previous state of data. LSTMs work based on a chain-based process in which 4 associated layers interact with each other. It also can be used for speech and music processing and other applications related to time-series data. Long short-term memory networks are a complex area of deep learning. It can be hard to get your hands around what the long short-term memory networks are, and how terms like bidirectional and sequence to sequence relate to the field. The success of long short-term memory networks is in their claim to be one of the first implements to overcome the technical problems and deliver on the promise of recurrent neural networks. In the recurrent neural networks, the exploding gradient and vanishing gradient are two major problems. Both of these problems related to how the network is trained. While in the long short-term memory networks these two technical problems are solved. Therefore, the long short-term memory networks can solve numerous tasks not solvable by previous learning algorithms for recurrent neural networks.

The units of a long short-term memory network are used as building units for the layers of a recurrent neural network, often called a long short-term memory network. Long short-term memory networks enable recurrent neural networks to remember inputs over a long period of time. This is because long short-term memory networks contain information in a memory, much like the memory of a computer. The long short-term memory network can read, write and delete information from its memory. This memory can be seen as a gated cell, with gated meaning the cell decides whether or not to store or delete information, based on the importance it assigns to the information. The assigning of importance happens through weights, which are also learned by the algorithm. This simply means that it learns over time what information is important and what is not. In a long short-term memory network, you have three gates which are input, forget and output gate. These gates determine whether or not to let new input in (input gate), delete the information because it is not important (forget gate), or let it impact the output at the current time step (output gate). The gates in a long short-term memory network are analog in the form of sigmoid function, meaning they range from zero to one. The fact that they are analog enables them to do backpropagation. The problematic issues of vanishing gradients are solved through LSTM because it keeps the gradients steep enough, which keeps the training relatively short and the accuracy high.

Generative adversarial networks

Generative adversarial networks (GANs) [78], first introduced by Goodfellow et al., are generative models that can produce synthetic data analogous to the real training data. They mainly consist of two network components: (1) Generator and (2) Discriminator. The generator is in charge of learning the distribution of training data and generating the synthetic data similar to the training set. The discriminator is responsible for distinguishing the generated fake data and real data. When the training phase is finished, the generator can produce many data samples similar to the training set.

On the other hand, the discriminator can distinguish the real data from other types of data that want to mislead the discriminator (e.g., anomaly detection). Therefore, GANs have two basic applications related to the discriminator and generator concepts. The applications of GANs are increasing progressively. For example, game developers use the GANs to increase the resolution.

Let's explain the basic concept of the GANs by an example. You can consider the GANs as a game between two players. One of these players is a generator network while the other one is a discriminator network. The duty of the first player or the generator network has to create new samples that intend to come from the same distribution of the input data or training dataset. However, the second player or discriminator network has to examine samples to determine which of these data are real or fake. In other words, the discriminator network must discriminate the real samples from the fake ones while the generator network must generate new fake samples. Generally, the discriminator network trains using the traditional binary supervised learning classifier. You can also interpret that the generator network goal is to fool the discriminator network. For example, you can think of the generator network as a counterfeiter and the discriminator network as police. The task of the counterfeiter is to make fake money that is indistinguishable from real money so that fool the police while the police should learn how to distinguish the fake money and real money.

Radial basis function networks (RBFNs)

RBFNs [79] are particular kinds of deep learning algorithms that employ the radial basis functions for activation functions of units. Analogous to other DL algorithms, The RBFNs consist of input, output, and hidden layers. They are employed for the tasks such as classification, regression, and time-series applications. The RBFNs have the following characteristics:

- The radial basis functions perform the classification tasks by calculating the similarity of inputs to training data instances.
- The radial basis functions consist of the input vector, which presents the input of the model.
- This function explores the sum of weighted inputs, and the output is the same as classification problems (e.g., one unit per class).
- The units in the hidden layer have a Gaussian-like function, which converts the distances to the gaussian distribution.
- The output of the model is the same as typical neural networks (e.g., linear combination and sum of the inputs.

Multi-layer perceptrons (MLP)

Multi-layer perceptrons (MLPs) [80] are the first and essential concept of deep learning algorithms that should be learnt. MLPs are a part of the artificial neural networks,consisting of several layers of perceptrons with activation functions. MLPs have three main layers,namely the input layer, output layer, and output layer. They may have several hidden layers between input and output layers; however, the number of input and output layers is the same and equals MLPs also have many applications such as image processing, speech recognition, fingerprint-based localization, and so on.

Deep belief networks

Deep belief networks (DBNs) [81] are a branch of deep learning algorithms used to generate data and have several layers of latent and stochastic variables. The latent variables are made by binary numbers, which are named hidden nodes. DBNs are a collection of RBMs interconnected among the layers such that each RBM is placed between two other RBMs. DBNs have many applications especially related to computer vision tasks such as image and video processing.

Restricted Boltzmann machines

Restricted Boltzmann Machines (RBMs) [82] is a class of probabilistic-based artificial neural networks that can learn the distribution of a given set of data. They have many applications such as image processing, speech recognition, feature learning, modeling of motion style, quantum physics. RBMs have two main layers: visible and hidden nodes. The visible nodes are associated with the hidden nodes. RBMs also consist of a bias node, which is associated with the visible and hidden nodes.

Autoencoders

Autoencoders [83] are a particular kind of artificial neural network whose inputs and outputs are the same. They are mostly utilized for unsupervised learning tasks. A simple autoencoder consists of three layers (an input layer, a hidden layer, and an output layer). The input layers are encoded to the hidden layer, and the autoencoder tries to decode the hidden units to the input samples on the output layer. It can be used for unsupervised learning tasks and can be utilized for compression tasks such as image compression.

The autoecoders compress the input into a lower dimensional features and then reconstruct the output from this representation. The feature is a compact summary or compression of the input, also called the latent-space representation. An autoencoder consists of 3 components which are encoder, code and decoder. The encoder compresses the input and produces the code, the decoder then reconstructs the input only using this code. Therefore, to build an autoencoder you need an encoding method, a decoding method, and a loss function to compare the output with the target. Autoencoders are mainly a dimensionality reduction or compression algorithm with a couple of important properties. The first property is that they are data specific. In other words, the autoencoders are only able to meaningfully compress data similar to what they have been trained on. Since they learn features specific for the given training data, they are different than a standard data compression algorithm like gzip. Therefore, you cannot expect an autoencoder trained on handwritten digits to compress landscape photos. The second property of autoencoders is that they are lossy. The output of the autoencoder will not be exactly the same as the input, it will be a close but degraded representation. If you want lossless compression they are not the way to go. Remember that these networks are classify to unsupervised category of machine learning algorithm. To train an autoencoder you do not need to do anything fancy and just throw the raw input data at it. Autoencoders are considered an unsupervised learning technique since they don't need explicit labels to train on. But to be more precise they are self-supervised because they generate their own labels from the training data.

Let's explore the details of the encoder, code and decoder. Both the encoder and decoder are fully connected feed forward neural networks. Code is a single layer of an artificial neural networks with the dimensionality of our choice. The number of nodes or neurons in the code layer (code size) is a hyperparameter that we set before training the autoencoder. First the input passes through the encoder, which is a fully connected artificial neural network, to produce the code. The decoder, which has the similar artificial neural network structure, then produces the output only using the code. The goal is to get an output identical with the input. Note that the decoder architecture is the mirror image of the encoder. This is not a requirement but it's typically the case. The only requirement is the dimensionality of the input and output needs to be the same. Anything in the middle can be played with. There are 4 hyperparameters in autoencoders and should be set before training an autoencoder. The first hyperparameter is code size. This refers to number of nodes or the neuron in the middle layer. Smaller size results in more compression. The second hyperparameter is the number of layers. The autoencoder can be as deep as we like. The third hyperparameter is the number of nodes or neurons per layer. The autoencoder architecture is called a stacked autoencoder since the layers are stacked one after another. Usually stacked autoencoders look like a sandwich. The number of nodes per layer decreases with each subsequent layer of the encoder, and increases back in the decoder. Also the decoder is symmetric to the encoder in terms of layer structure. As noted above this is not necessary and we have total control over these parameters. The number of neurons for each layer and the number of layers hyperparameters are similar hyperparameters when someone designing an artificial neural network. Remember that if the network is deep and number of neurons for each layer is lager, the autoencoder can face with overfitting problem. Therefore, you should use the regularization technique to handle the overfitting phenomena. The third part of an autoencoder is loss function. There are many loss functions like the mean squared error (mse) or binary crossentropy. If the input values are in the range between 0 and 1 then it is better to use crossentropy, otherwise the mean squared error can be used. To train an autoencoders you can use the backpropagation like the artificial neural networks.

Challenges

Like many other machine learning algorithms, ANNs also have some challenges, such as overfitting problems and computations in the training and test phases. DNNs incline to an overfitting problem since adding each layer increases the power of a DNN model to fit data very well. Researchers have introduced several approaches to solve the overfitting problem, such as ℓ_2-regularization and ℓ_1-regularization [70]. In another instance, the dropout layers [71] can be used to handle this problem. The dropout layers vanish some units of layers with a user-defined probability. Alternatively, other straightforward ways can be used to diminish the overfitting chance, such as augmenting the data by cropping or rotating them into different directions. Another important issue is the time computation of DNNs. Several parameters and options must be regarded for the DNNs, such as the number of layers, number of nodes for each layer, rate of learning, and initialization of weights. Finding the best parameters maybe not be practical, and it is like finding a needle in a haystack. Several tricks are used to tackle these problems, such as batching [72]. Also, researchers can utilize other types of artificial neural networks that are used to speed up the computation. For example, deep extreme learning machines have a faster training phase than other algorithms [73, 74].

CHAPTER 6

Generative Adversarial Networks

Machine learning algorithms are mainly divided into supervised, semi-supervised, and unsupervised algorithms. The supervised algorithms need labeled data for the optimization process. However, the data collection phase for these algorithms is very hard. Many algorithms have been introduced in recent years to reduce human-centered data collection, which is very expensive process. The semi-supervised and unsupervised algorithms have mainly focused on reducing cost. Specifically, the variational autoencoder (VAE) and Generative adversarial networks (GANs) can produce artificial data samples based on a limited set of the collected data. The VAE can produce the data. However, in computer vision applications, the generated images are blurry. Generative adversarial networks, which are semi-supervised learning algorithms, generally have to generate the synthetic data with a small set of collected data. The generated data is not the same as the collected data; however, they are very similar. This data can be used for supervised machine learning algorithms as an additional input to increase up their accuracy. GANs are amongst the foremost essential research topic within different research fields including, image-to-image translation [84], fingerprint localization [59], hyperspectral classification [85], speech and language processing [86, 87], malware detection [88] and video generation and prediction [89, 90]. First, in this section, the conventional model of GANs is introduced, and then the other evolved models from the main structure of GANs are presented.

Generative Adversarial Networks (GANs)

The first model of GANs was introduced by Goodfellow et al. in 2014 [78]. This model mainly consists of two Generator (G) and Discriminator (D) nets. First, we describe the training phase of the conventional GANs, which is depicted in Figure 6.1. The noise is the input of the G net, and the real data is its output. The goal of the G net is to convert the noise to the data. Then, the D net learns the distribution of real/fake data and increases its ability. The learning phase of these two nets are performed based on the following cost function

$$\min_{G} \max_{D} \mathcal{L}(D,G) = E_{x \sim p_r(x)} \left[\log D(x)\right] + E_{z \sim p_z(z)} \left[\log(1 - D(G(z)))\right] \tag{1}$$

The above equation has two main parts, and the goal is to give true labels to the real and fake data as much as possible. The G and D nets are two functions, which in the context of deep learning means that they can be denoted by multilayer perceptron (MLP). The G net trains to convert the noise $z \sim p_z(z)$ to real data. This net can be indicated by $G(z; \theta_g)$, in which θ_g is the parameters in the G net. Then, the D net trains to discriminate between real and fake data. This net can be indicated by $D(z; \theta_d)$, in which θ_d is the parameters of the D net. The structure of this net is binary classification, in which zero and one show the fake and real collected data, respectively. These two nets can be defined by separate cost functions. For D net, the cost function is as follows

$$\mathcal{L}(\theta_d) = E_{x \sim p_r(x)} \left[\log D(x; \theta_d)\right] + E_{z \sim p_z(z)} \left[\log(1 - D(G(z; \theta_g)))\right] \tag{2}$$

And the cost function for the G net is as follows

$$\mathcal{L}(\theta_g) = E_{z \sim p_z(z)} \left[\log(1 - D(G(z; \theta_g)))\right] \tag{3}$$

The θ_d and θ_g are updated step by step with algorithm 1. When the D net cannot discriminate between real and fake data, convergence occurs. In other words, when $D(x; \theta_d) = 0.5$ the convergence occurs and these two nets are ready. Afterward, the G net is ready to produce artificial data from the noise $z \sim p_z(z)$.

First, the needed parameters are initiated in Algorithm 1. There are two loops for updating the parameters of the G and D nets. Batch samples from the noise and real data samples are picked up in the inner loop. Then, the

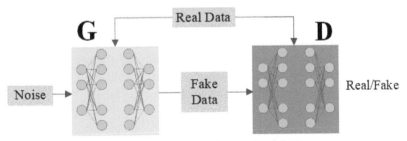

Figure 6.1. Schematic of conventional GAN.

loss of the D net is calculated based on the given sample batches. Afterward, the parameters of the D net are updated based on the given rate, and the same process is performed for the G net. An important note is that the D net parameters are updated I times larger than the G net.

Algorithm 1 (Training of G nad D in GAN)

---

initialization $\begin{cases} O = \text{number of epochs, } I = \text{Iteration of inner loop} \\ \text{updating rates}\,(\alpha_g \text{ and } \alpha_d) \end{cases}$

for (i=0; i < O; i++){

    for (j=0; i < I; i++) {

        batch samples from noise $\mathbf{Z} \in \mathbb{R}^{B \times L} \sim p_z(z)$

        batch samples from data $\mathbf{X} \in \mathbb{R}^{B \times M}$

$$\mathcal{L}(\theta_d) = \frac{1}{B}\sum_{b=1}^{B}\Big[\log D(\mathbf{X}_b,\theta_d) + \log(1 - D(G(\mathbf{Z}_b,\theta_g)))\Big]$$

$$\xi_d = \frac{\partial}{\partial \theta_d}\mathcal{L}(\theta_d)$$

$$\theta_d^{j+1} = \theta_d^t + \alpha_d \xi_d$$

    }

    batch samples from noise $\mathbf{Z} \in \mathbb{R}^{B \times L} \sim p_z(z)$

$$\mathcal{L}(\theta_g) = \frac{1}{B}\sum_{b=1}^{B}\log(1 - D(G(\mathbf{Z}_b,\theta_g)))$$

$$\xi_g = \frac{\partial}{\partial \theta_g}\mathcal{L}(\theta_g)$$

$$\theta_g^{i+1} = \theta_g^i - \alpha_g \xi_g$$

}

In Figure 6.2, the digits are generated by feeding the noise samples to the G net based on the figures of the MNIST dataset. As can be seen below, by increasing the number of epochs, the quality of figures improves until reaching a certain level. Also, the noise generates arbitrary digits, and we do not have control for generating the images with class labels.

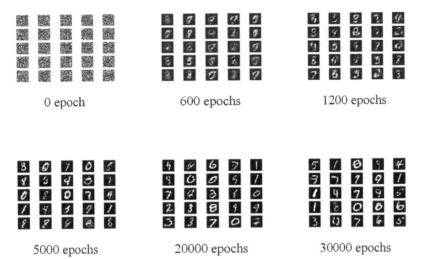

Figure 6.2. Generated images by conventional GAN after a certain number of epochs.

Conditional GAN (CGAN)

The conventional GAN can only produce the data without class labels. However, we need the class labels of the generated samples. In this case, the easiest way is to train different models for each class separately. However, it consumes so many costs. Therefore, Mirza and Osindero [91] suggested another model, named Conditional GAN (CGAN), to solve this problem, which uses encoded class labels for the inputs of the G and D nets to produce the samples conditioned on the class labels, as depicted in Figure 6.3. The cost function is the same as the conventional GAN by performing a subtle difference in the G and D nets, in which inputs are conditioned on the class labels as follows

$$\min_{G} \max_{D} \mathcal{L}(D,G) = E_{x \sim p_r(x)} \left[\log D(x \mid y) \right] + E_{z \sim p_z(z)} \left[\log(1 - D(G(z \mid y))) \right] \quad (4)$$

The D net has the following cost function

$$\mathcal{L}(\theta_d) = E_{x \sim p_r(x)} \left[\log D(x \mid y; \theta_d) \right] + E_{z \sim p_z(z)} \left[\log(1 - D(G(z \mid y; \theta_g))) \right] \quad (5)$$

And the G net cost function is as follows

$$\mathcal{L}(\theta_g) = E_{z \sim p_z(z)} \left[\log(1 - D(G(z \mid y; \theta_g))) \right] \quad (6)$$

In Figure 6.4, the generated images by CGAN have been plotted. It can be seen that the CGAN needs more epochs for generating high-quality data. However, it can produce the images conditioned on the class labels, allowing us to generate the data with class labels.

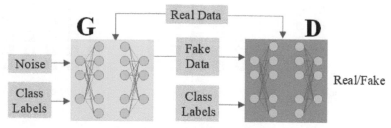

Figure 6.3. Conditional GAN (CGAN).

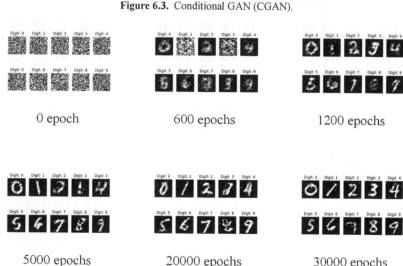

Figure 6.4. Generated images by CGAN after a certain number of epochs.

Auxiliary Classifier GAN (AC-GAN)

As mentioned, a conventional GAN is not able to produce data with class labels. Odena et al. [92] suggested another model for producing the labels for the generated data, named Auxiliary Classifier GAN (AC-GAN). The structure of this model is depicted in Figure 6.5. In the AC-GAN, the class label $c \sim p_c(c)$ beside the latent noise $z \sim p_z(z)$ is fed to the G net. The G net is conditioned on the class label similar to the G net in the CGAN. There a multitask MLP consists of two D and Q nets, where the generated data are fed to the input of this multitask MLP. One task is to recognize the real and fake data, and the other one is to produce the class labels. The cost function for this model can be represented by two separate functions as follows

$$\mathcal{L}_S = E_{x \sim p_r(x)}\left[\log D(x)\right] + E_{z \sim p_z(z)}\left[\log(1 - D(G(z)))\right] \tag{7}$$

$$\mathcal{L}_C = E_{x \sim p_r(x)}\left[\log Q(c \mid x)\right] + E_{z \sim p_z(z)}\left[\log(Q(c \mid z)\right]$$

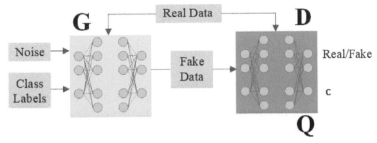

Figure 6.5. Auxiliary classifier GAN (AC-GAN).

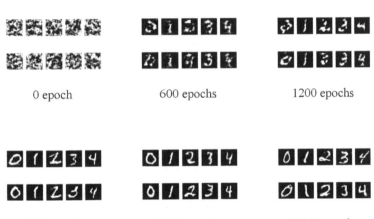

Figure 6.6. Generated images by ACGAN after a certain number of epochs.

The G net maximizes the $\mathcal{L}_S + \mathcal{L}_C$, and the D net maximizes the $\mathcal{L}_S - \mathcal{L}_C$ till reaching a convergence.

In Figure 6.6, the generated images by AC-GAN have been plotted. It can be seen that the AC-GAN can produce data with class labels and also needs fewer epochs for generating high-quality data compared with CGAN.

Wasserstein GAN (WGAN)

Gradient vanishing is one of the main concerns of conventional GAN. Therefore, the training phase is not so easy. Several algorithms have been suggested in the past few years to solve this problem. Arjovsky et al. [93] first suggested adding extra noise to the generated data arrived from the G net before moving the generated data to the D net. In the other work, Arjovsky et al. [94] suggested a new algorithm with a new cost function, named Wasserstein GAN

(WGAN), to solve the mentioned bottleneck. The introduced cost function is as follows:

$$\mathcal{L}(F,G) = \sup_{\|F\|_{L} \leq 1} E_{x \sim p_r(x)} \big[F(x) \big] - E_{z \sim p_z(z)} \big[F(G(z)) \big] \tag{8}$$

where sup represents the supremum.

In Figure 6.8, the generated images by WGAN have been plotted based on a certain number of epochs. It seems that the WGAN needs more epochs compared with conventional GAN for generating the data, and some figures are not similar to the real digits. The presented problem is the main concern of the WGAN, which occurs due to the undesired behavior of wright clipping.

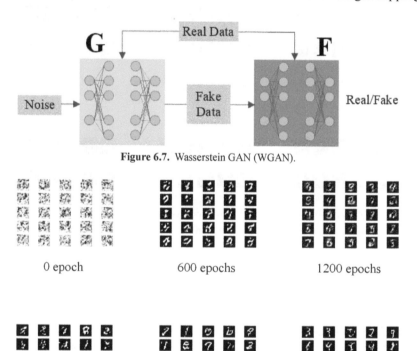

Figure 6.7. Wasserstein GAN (WGAN).

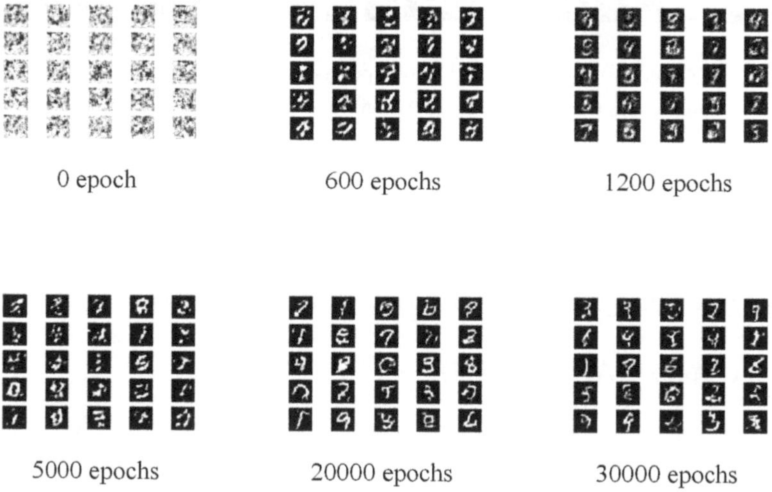

0 epoch   600 epochs   1200 epochs

5000 epochs   20000 epochs   30000 epochs

Figure 6.8. Generated images by WGAN after a certain number of epochs.

WGAN with Gradient Penalty (WGAN-GP)

As mentioned, the conventional GAN suffers from instability problems, and WGAN enhances the stability of the conventional GAN during the training phase. However, WGAN also suffers from the unpleasant treatment of the critic weight clipping during the training phase. To solve this problem, Ishaan Gulrajani et al. [95] suggested a new cost function, named WGAN with Gradient Penalty (WGAN-GP). They used a gradient penalty for the cost function of WGAN, which is as follows

$$\mathcal{L}(F,G) = \sup_{\|F\|_{L} \leq 1} E_{x \sim p_{r}(x)} \left[F(x) \right] - E_{z \sim p_{z}(z)} \left[F(\underbrace{G(z)}_{\tilde{x}}) \right] + \beta E_{\hat{x} \sim p_{\hat{x}}(\hat{x})} \left[\left(\| \nabla_{\hat{x}} F(\hat{x}) \|_{2} - 1 \right)^{2} \right] \quad (9)$$

where $\hat{x} = \gamma \overline{x} + (1 - \gamma)x$ and $0 \leq \gamma \leq 1$. In Figure 6.9, we have plotted the new digits produced by WGAN-G. As can be seen, the WGAN-GP can produce the digit images with fewer epochs, and also, the quality of figures is better compared with WGAN.

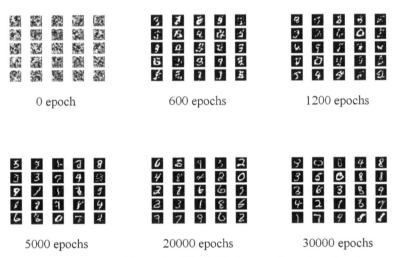

Figure 6.9. Generated images by WGAN-GP after a certain number of epochs.

Info GAN

Xi Chen et al. [96] suggested a new GAN learning algorithm, named Info GAN, for producing the data with extra information instead of using an arbitrary noise to the G net's input. For instance, the extra information can be the angles or thickness of the digits' pictures next to the noise when producing the data for the MNIST dataset. The latent variables indicate this extra information are c_1, c_2, ..., c_L. We can suppose that these variables do not have correlated information, and they are independent. For example, the thickness and angle information are not correlated, and this assumption is not an impossible condition. Therefore, the joint probability can be written as $P(c_1, c_2, ..., c_3) = \prod_{i=1}^{L} P(c_i)$, where c is a vector which consists of all c_i variables, and the cost function is defined as follows

$$\min_{G} \max_{D} \mathcal{L}_{Info}(D,G) = \mathcal{L}(D,G) - \beta I(c; G(z,c)) \tag{10}$$

where $\mathcal{L}(D, G)$ equals to $E_{x \sim p_r(x)}[\log D(x)] + E_{z \sim p_z(z)}[\log(1 - D(G(z)))]$, the $I(c;$ $G(z, c))$ equals to $H(c|G(z, c))$, $H(c)$ is the entropy of latent variables and $H(c|G(z, c))$ is the entropy of latent variables conditioned on the G net data. We cannot directly maximize the $I(c; G(z, c))$, since we should attain the probability distribution of latent variables conditioned on the inputs, which is denoted by $P(c|x)$. Nevertheless, we can attain its lower bound by a new net architecture denoted by $Q(c|x)$, which is known as auxiliary distribution.

$$I(c; G(z,c)) \geq E_{c \sim p(c), x \sim G(z,c)}\left[\log Q(c|x)\right] + H(c) = E_{x \sim G(z,c)}\left[E_{c' \sim p(c|x)}\left[\log Q(c'|x)\right]\right] + H(c)$$
$$= \mathcal{L}_I(G,Q) \tag{11}$$

therefore, the final cost function derived from (9) and (10) can be defined as follows:

$$\min_{G,Q} \max_{D} \mathcal{L}_{Info}(D,G,Q) = \mathcal{L}(D,G) - \beta \mathcal{L}_I(G,Q) \tag{12}$$

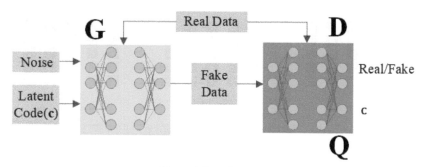

Figure 6.10. Info GAN.

Least Square GAN (LSGAN)

As mentioned, gradient vanishing is one of the main concerns in conventional GAN. Xudong Mao et al. [97] proposed a novel cost function to deal with this problem of GAN during the training phase. Their proposed model is similar to conventional GAN unless they replaced the cross-entropy by the least-squares cost function, as depicted in Figure 6.1. For the D net, the cost function can be written as follows

$$\min_{D} \mathcal{L}(D) = E_{x \sim p_r(x)}\left[(D(x) - 1)^2\right] + E_{z \sim p_z(z)}\left[(D(G(z)) - 1)^2\right] \tag{13}$$

And for the G net, the cost function can be defined as follows

$$\min_G \mathcal{L}(G) = E_{z \sim p_z(z)}\left[(D(G(z)) - 1)^2 \right] \tag{14}$$

In Figure 6.11, the generated images by the LSGAN have been plotted. It seems that the LSGAN needs more epochs for producing high-quality data compared with the conventional GAN.

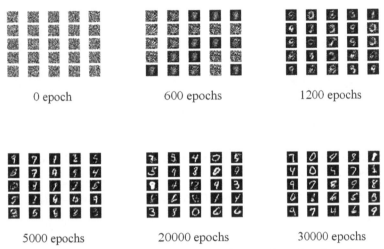

0 epoch   600 epochs   1200 epochs

5000 epochs   20000 epochs   30000 epochs

Figure 6.11. Generated images by LSGAN after a certain number of epochs.

Bidirectional GAN (BiGAN)

The conventional GAN converts the noise to the real data; nevertheless, it cannot invert the real data to the latent noise. Donahue et al. [98] suggested a new unsupervised GAN algorithm, named Bidirectional GAN (BiGAN), depicted in Figure 6.12. This new model converts the real data to the latent

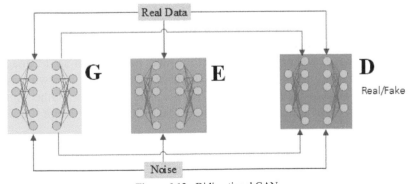

Figure 6.12. Bidirectional GAN.

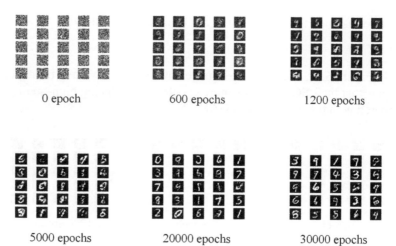

Figure 6.13. Generated images by BiGAN after a certain number of epochs.

noise with a new Encoder net (E) next to the G and D nets. The cost function is defined as follows

$$\min_{G,E} \max_{D} \mathcal{L}(D,E,G) = E_{x \sim p_r(x)} \Big[E_{z \sim p_E(\cdot|x)} \log D(x,z) \Big] + E_{z \sim p_z(z)} \Big[E_{x \sim p_G(\cdot|x)} \log(1 - D(G(x,z)) \Big] \, (15)$$

In Figure 6.13, the generated digit images by BiGAN have been illustrated. As can be seen below, BiGAN needs more epochs; however, it can generate high-quality data by increasing the number of training epochs.

Dual GAN

As mentioned, the conditional GAN can produce the data conditioned on the class labels. However, we cannot access the class labels for many applications since human-centered labeling is expensive, and so many pair labels are needed. Yi et al. [84] proposed a new GAN-based learning framework motivated by natural language translation from a dual perspective (e.g., English-to-French and vice versa). The general purpose of this model is the image-to-image translation, which works based on two sets of unlabeled data (e.g., photo and sketch). This model consists of a G and Dpair nets for one set of images and another G and Dpair nets for the other set of images, in which set 1 is translated to set 2. Assuming that U and V are two sets of images, the goal of G_A to learn to map an image $u \in U$ to another image $v \in V$. At the same time, the goal G_B is to learn to map an image $v \in V$ to

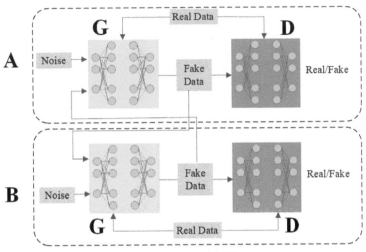

Figure 6.14. Dual GAN.

another image $u \in U$. This model is depicted in Figure 6.14, and the cost function for D nets are defined as follows:

$$
\mathcal{L}_A^D(u,v) = D_A(G_A(u,z)) - D_A(v)
$$
$$
\mathcal{L}_B^D(u,v) = D_B(G_B(u,z')) - D_B(v)
$$

(16)

where the noise of the G net is for the A and B parts, respectively. Designers of dual GAN proposed using dropout layers for both of the test and train phases instead of feeding arbitrary noise to the input of generators. The same cost function is used to define the generators G_A and G_B.

Deep Convolutional GAN (DCGAN)

Conventional GAN uses simple dense layers for the G and D nets. In the past few years, the convolutional layers have been successfully used for supervised machine learning algorithms. However, in the context of semi-supervised and unsupervised learning, it has gained less attention. Several studies show that the convolutional layer has better performance for image datasets. Radford et al. [99] suggested using deep convolutional layers for the G and D nets. Their proposed method produces high-quality data compared with conventional GAN. In Figure 6.15 the generated images by DCGAN have been plotted by a certain number of epochs. As can be seen, the generated images are more natural than the dense layers, and the quality of figures is better.

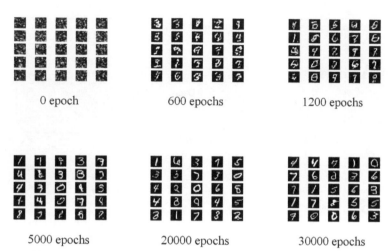

0 epoch   600 epochs   1200 epochs

5000 epochs   20000 epochs   30000 epochs

Figure 6.15. Generated images by DCGAN after a certain number of epochs.

CHAPTER 7

Implementation

||||||||||||||||||||||||||||||||||||||||||||||||||||||||||||||||||||||||||||||||||||||||||||||||||||||||||||||||||||||||||||||||||||||||||||||||||||||||||||

The blended effect of new computing resources and techniques in addition to the expansion of large datasets is reshaping many research areas. It may result in technological innovations usable by billions of people. In recent years, Machine Learning and particularly its subareas, Deep Learning, have seen noble improvements. Techniques elaborated within these two fields can dissect and get the hang of massive amounts of real-world examples in diverse formats. The number of Machine Learning algorithms is broad and growing and their implementations through frameworks and libraries. Software development in this field is rapid, with much open-source software coming from open-source communities. This section presents a recent time-slide inclusive overview with collations and trends in the development and usage of brand-new Artificial Intelligence software. It also provides an overview of huge parallelism support capable of scaling computation effectively and efficiently in the epoch of Big Data.

Applying a machine learning technique to a given set of data, depending on the computing and storage capabilities available for the scientist, may take a considerable amount of time. In this context, different types of accelerators capable of computing specific tasks have been successfully used to complement and unburden more generic CPUs in many areas, e.g., arithmetic co-processors for floating-point operations, sound cards for audio encoding/decoding, or 3D graphics accelerators for image-array operations and conversions. Recently, a rise of new type of accelerators like cryptographic and AI accelerators (such as Google's TPUs [100]) has emerged, resulting from new hardware architecture generation optimized for ML DL, and other AI workloads. Generally, in the last decade, the graphics processing units (GPUs) have become the main tools to speed up general-purpose computation [101]. They offer massive parallelism to extend algorithms to large-scale data for a fraction of the cost of a traditional high-performance CPU clustering, allowing scalability over datasets that are not computable through traditional parallel approaches.

Machine Learning techniques rely on the analysis of data for getting insights into it, which provides relevant information for the problem that is being analyzed. Here and now, Big Data has permeated all disciplines and research areas (such as computer science, medicine, finance, etc.) due to its potential within all these fields. The generation and collection of data has been modified,leading to changes in the data processing. Volume, velocity, variety, value, and veracity characterize the definition of Big Data. At first glance, the general audience relates Big Data processing with distributed platforms such as Apache Hadoop or spark. The same thing goes there for volume in respect to veracity characteristic [102] and the high-speeds in processing and inference in background. Data analysis is anticipated to evolve in the new era of Big Data. New approaches and tools that can accommodate data with different structures, different spatial and temporal scales [103] are required for the nature of large-scale data. The burst of the large volume of information, with the characteristic variety in particular and processed by data mining and Machine Learning algorithms, requires a new transformative distributed and parallel computing solutions potent of scaling computation effectively and efficiently [101]. In this context, a comprehensive overview is provided with comparisons beside trends in the development and usage of fresh AI software, libraries, and frameworks able to learn to adapt to previous experience using Machine Learning and DL techniques to perform more truthful and more functional operations for problem-solving.

Accelerated computing

DL takes advantage of using a form of specialized hardware extant in accelerated computing environments. The current conventional regnant solution (NVIDIA) uses Graphics Processing Units (GPUs) as general-purpose processors. GPUs supply immense parallelism for large-scale DL problems, making it possible to scale algorithms vertically to huge amounts of data that are not traditionally computable [101]. GPUs are advantageous solutions for real-world and real-time systems demanding expeditious decision-making and learning (especially in image processing). Field Programmable Gate Array (FPGA) [104] and the lately made public Google Tensor Processing Unit 3.0 (TPU) are alternatives to GPUs. Other IT companies are also proceeding to offer dedicated hardware for DL acceleration, e.g., Kalray with their second generation of DL acceleration device named MPAA2-256 Bostan, focused on mobile devices like autonomous cars.

There are many other DL accelerator technology providers, including worldwide IT companies such as IBM TrueNorth Neuromorphic chip,

Microsoft Brain Wave, Intel Nervana Neural Network Processor, AMD Radeon Instinct, as well as many startups. This hardware needs to be accommodated with the new trends emerging in the ML/DL software development community. For example, several schemes that will immensely profit from specialized hardware have been designed to improve the speed and memory consumption of DL algorithms:

- *Sparse Computation*, which forces the use of sparse representations along with the neural network. Lower memory requirements and faster computation are the benefits.

- *Low precision* data types, smaller than 32-bits (e.g., half-precision or integer), with examination even by 1-bit computation [105]. Likewise, this speeds up algebra calculations besides decreasing memory consumption at the cost of a rather less accurate model [106]. In the last few years, 16-bit and 8-bit computations are starting to be supported by most DL frameworks. The vertical scalability of large-scale DL is still limited because of the GPU memory capacity, which is 32 GiB at the maximum on the NVIDIA Volta architecture. Integration of the MapReduce frameworks with GPU computing may outdo many of the performance confinements and manifest a challenge for future research. A combination of hardware resources to scale out to bigger data (data parallelism) or bigger models (model parallelism) is feasible using Multi-GPU and distributed-GPU solutions. Data parallelism and model parallelism are different ways of distributing an algorithm and are often used in the context of Machine Learning on how make accessible the computing power to facilitate computations [107].

- Data parallelism engages the use of different nodes to run the same proportion of code on different batches of data.

- Model parallelism brings in the development of more complex models distributing the computation of different model subparts within different worker nodes. Simultaneously, parallelization at the model level has one main restricting factor; the communication latency between cluster nodes, which is abatable using specialized interconnections. Krizhevsky (2014) provides advantages (and disadvantages) of combining data parallelism (in the convolutional layers) and model parallelism (in the dense layers) of CNN [108].

Many-core accelerators such as GPU have the main feature of intensely parallel architecture, which allows computations involving matrix-based operations, which are at the core of many ML/DL implementations to be hastened. Manufacturers often presume the possibility of enhancing hardware configuration with many-core accelerators to enhance machine/cluster performance in addition to accelerated libraries, which gives highly optimized primitives, algorithms, and functions to have the massive parallel ability of GPUs at hand. The most common accelerated libraries include:

- The NVIDIA CUDA (Compute Unified Device Architecture) is a parallel computing platform and programming model developed by NVIDIA for general computing on GPUs. GPU-accelerated CUDA makes drop-in acceleration across multiple domains such as linear algebra, image/video processing, Deep Learning, and graph analytics possible. The NVIDIA CUDA Toolkit delivers a development environment for creating high-performance GPU-accelerated applications.

- The NVIDIA CUDA Deep Neural Network library (cuDNN), is a GPU-accelerated library of DNN's primitives. The cuDNN offers highly tuned implementations for standard routines, including pooling, activation layers, normalization, and forward/backward convolution. Using cuDNN, instead of struggling with low-level GPU performance tuning, Deep Learning users can concentrate on training Neural Networks and developing software applications. The cuDNN is used by various DL frameworks such as PyTorch, MatLab, CNTK, TensorFlow, Theano, and Caffe2.

- Intel MKL (Intel Math Kernel Library) optimizes code with as little effort as possible for future generations of Intel processors. It is compatible with collecting operating systems, languages, linking/threading models, and compilers. By accelerating math processing routines, it increases application performance while reducing development time. This accessible core math library includes Fast Fourier Transforms (FFT), vector statistics and data fitting, vector math, miscellaneous solvers, sparse solvers, Deep Neural Networks, and linear algebra (BLAS/OpenBLAS, LAPACK, ScaLAPAK).

- OpenCL (Open Computing Language) hands over compatibility across heterogeneous hardware from any supplier. As a prospective forward lane, the AMD ROCm (Radeon Open Compute Platform) open ecosystem is mentioned here. ROCm is AMD's response to the prevalent NVIDIA CUDA and involves a programming-language independent open-source HPC/hyper scale-class platform for GPU computing. It spotlights:

- Peer-to-peer multi-GPU operations beside Remote Direct Memory Access (RDMA) support.

- The Heterogeneous System Architecture (HAS) runtime API offers a rich foundation to execute C++, Python, OpenCL, HIP, and HCC.

- The Heterogeneous-compute Interface for Portability (HIP) allowing developers to convert CUDA code to conventional C++. The HIP is very light and hardly impacts the performance of coding directly in CUDA or HCC. As CUDA and HIP are C++ languages, porting from CUDA to HIP is conspicuously simpler than porting from CUDA to OpenCL.

- MIOpen, an AMD GPU-accelerated open-source library for high-performance Machine Learning primitives with large cuDNN compatible portions of source code. At this time, MIOpen only supports Caffe, Caffe2, and TensorFlow. Other Deep Learning libraries such as HIPnn, PyTorch, CNTK, and MXNet are on the development list. Other programming libraries supporting computational speed-up and parallel computing are:

- OpenMP, an Application Programming Interface (API) supporting multi-platform shared memory multiprocessing programming, including a set of environment variables, compiler directives, and library routines effective on run-time activities.

- OpenMPI, an open-source implementation of the MPI specifications. The OpenMPI software obtains high performance and is well susceptible to community input. MPI stands for the Message Passing Interface, a regulated API, mostly used for parallel or distributed computing. The MPI forum, a big committee dealing with a cross-section between research agents and the industry, has written the MPI. Applications developed with the hybrid model of parallel programming can run on clusters using both MPI and OpenMP; OpenMP is used for parallelism within a (multi-core) node and the use of MPI for parallelism between the nodes.

Machine learning frameworks and libraries

There are many Machine Learning algorithms together with their different software implementations. Many software tools associated with Machine Learning techniques have been worked on for the last few years [109]. Their mutual intent is to speed up the complex data analysis process and propose integrated environments up-to-date with standard programming languages. These tools have been developed for different objectives such as processors (for images, sound, or language), analytics platforms, recommender systems, and predictive systems.

Once more, it is important to point out that there is no single tool fitting every problem, and most of the time, a blend of them is needed for success. That the code of many open-source tools is hosted on GitHub in the shape of repositories is a fact. GitHub also retains a lot of monitoring information about software development, namely the number of contributors and commits (besides historical and ongoing activities of each member of the team and the project altogether), number of issues, stars, forks, and watches, along with insights and diagrams. Many third-party applications connected to GitHub offer automated code analytics and reviews which their qualities and quantities are all based on GitHub data. However, they differ in presentations, viewpoints, evaluation details, and preference settings.

No need for special hardware support

Shogun

Shogun, an adept open-source general-purpose Machine Learning Library, offers a vast range of resourceful and combined Machine Learning methods [110] built on an architecture developed in C++ and licensed GNU GPLv3 license and has been under-engaged in development since 1999.

At the moment, a team of various volunteers develops Shogun. It is a sponsored project of NumFOCUS since 2017. The key concept behind Shogun is that the implicit algorithms are transparent and at hand, which means anyone can use them for free. There are 15 implementations mixed with more than 35 kernel implementations in the SVM library, which are more combinable/constructible using sub-kernel weighting. Shogun covers a vast range of regression and classification methods in addition to several algorithms to train hidden Markov models, linear methods, statistical testing, distance counting, model evaluations, FFNNs, clustering, and many more. Shogun has been successfully put into practice with computer vision, stock market analysis, speech and handwriting recognition, medical diagnosis, invasion detection bioinformatics, network security, object recognition, and many more. It is transparently usable in many languages and environments, namely Ruby, C#, Python R, Java, Scala, Lua, Octave, and R.

Strengths

- Breath-oriented Machine Learning toolbox consisting of numerous standard and brand-new Machine Learning algorithms
- API-oriented, open-source, cross-platform, earliest yet maintained by core implementation using C++
- Bonds with many other Machine Leaning libraries and programming interfaces in a variety of languages.

Weaknesses

- Most of the code has been written by academics for their studies over a long time; thus, maintaining or extending the code is not simple.
- There is no documentation. Therefore it is mostly for academic use.

RapidMiner

RapidMiner is a general-purpose data science software platform designed for Machine Learning, Deep Learning, text mining, predictive analytics, and data preparation [111]. It is a cross-platform framework developed on an open core model written in java. The development of RapidMiner (also known as YALE, Yet Another Learning Environment) started in 2001 at the Artificial Intelligence Unit of the Technical University of Dortmund.

RapidMiner could be interactive and supports Graphical User Interfaces (GUIs), while the command-line interface (CLI) and Java APIs are also available. The architecture of RapidMiner is based on a client/server model. In contrast, the server is offered as either using public or private cloud infrastructures (Amazon AWS and Microsoft Azure) or on-premise. RapidMiner supports unsupervised learning in Hadoop (Radoop), supervised learning in memory with scoring on the cluster (SparkRM), and scoring with native algorithms on the cluster for large-scale data analytics. RapidMiner changes its model to *business source,* meaning the latest versions will always be available as trials or under a commercial license even though its core stays open-source. The free version is limited to one logical processor and 10,000 data rows, available under the AGPL license.

Strengths

- It is General purpose and includes a wide set of algorithms with learning schemes and models and algorithms from Weka and R scripts.
- Supports Add-ons with chosen algorithms for large-scale data.
- It is a cross-platform framework with a strong community

Weaknesses

- Exclusive product for bigger problem solutions

Scikit-Learn

Scikit-Learn is generally known as a leading open-source tool for Python in which there is an inclusive library for Machine Learning algorithms. David Cournapeau started it as a Google Summer of Code project. Following 2015, Scikit-Learn is under active development sponsored by Telecom ParisTech, INRIA, and sometimes Google through its Summer of Code. Scikit-Learn has broadened the functionality of SciPy and NumPy packages along with many Machine Learning algorithms while offering functions to perform regression, classification, clustering, model selection, preprocessing, and dimensionality reduction. Matplotlib package is also used by Scikit-Learn to plot charts. From April 2016, it is conveyed in together-developed Anaconda for Cloudera project on Hadoop clusters. Along with Scikit-Learn, Anaconda consists of many popular packages for science, mathematics, and engineering for the Python ecosystem, namely Pandas, SciPy and NumPy.

Strengths

- Popular Python tools, commercially usable, open-source, and general-purpose
- Sponsored by Telecom ParisTech, INRIA, Google, and others

- Up-to-date and inclusive set of algorithms and implementations
- A fraction of many ecosystems and closely blended with scientific and statistic Python packages

Weaknesses

- API-oriented only
- The library does not support-GPUs
- Only basic tools for Neural Networks are included.

LibSVM

LibSVM is a comprehensive library for Support Vector Machines (SVMs), and its development began in 2000 at National Taiwan University. This library is written C/C++ and partly Java. Learning duties of LibSVM include distribution estimation, support vector classification (SVC) for binary multi-class, and support vector regression (SVR). It supports C-SVC, v-SVC, ε-SVR, and v-SVR and distribution estimation (one-class SVM) formulations.

These formulations, which are quadratic minimization problems, can be solved using a sequential minimal optimization algorithm. LibSVM uses various penalty parameters in the SVM problem formulation, which allows it to offer novel settings for unbalanced data.

With 250K downloads from 2000 to 2010, It has been successful in practice with NLP, Neuro-imaging, bioinformatics, and computer vision. LibSVM provides interfaces for PHP, Python, Perl, Ruby, Common Lisp, Clips, LabVIEW, R, KNIME, RapidMiner, Weka, OCaml, and Haskell. Scikit-Learn modifies and improves LibSVM to use it for handling computations internally.

Strengths

- LibVSM has its specific data format for the data analysis tool LibVSM that is compatible with other libraries and frameworks. The format is solid and appropriate for describing and processing Big Data, particularly because it allows a sparse representation.
- It is a popular, well-designed tool as well as being open-source.

Weaknesses

- LibSVM training algorithm is not as practical as LibLinear or Vowpal Wabbit in scaling up large datasets. Its time complexity is $O(n^3)$ [112] in the worst case and about $O(n^2)$ common cases, where n is the number of data inputs.
It- Only beneficial to problems where SVM acts expeditiously.

LibLinear

LibLinear is a library mainly designed to solve large-scale linear classification problems written in C++. Its development began in 2007 at National Taiwan

University [113]. LibLinear supports logistic regression and linear SVM as Machine Learning tasks, and ℓ_2-regularized logistic regression, ℓ_2-loss and ℓ_1-loss linear SVMs as problem formulations. The method for ℓ_1-SVM and ℓ_2-SVM is a coordinate descent approach. LibLinear implements a trust region Newton method for linear regression (LR) and ℓ_2-SVM.

Moreover, it implements the one-vs-rest strategy and the Crammer & Singer method [114] for multi-class problems while providing interfaces for Java, Python, Octave, MatLab, and Ruby. Scikit-Learn modifies and improves LibLinear to use it for handling computations internally. The Machine Learning group at National Taiwan University also supports (in early stages) MPI LibLinear. It is an extension of LibLinear for distributed environments and Spark LibLinear, rooted in LibLinear and linked with the Hadoop distributed file system. LibLinear is a prominent library in the open-source Machine Learning community and is released under the 3-clause BSD license.

Strengths

- Designed for solving large-scale linear classification problems
- A popular, well-designed tool as well as being open-source

Weaknesses

- Confined to Linear Regression and linear SVM

Vowpal Wabbit

Vowpal Wabbit (VW) is a resourceful scalable implementation of online Machine Learning while supporting different incremental Machine Learning approaches. It is an open-source rapid learning system originally developed by John Langford at Yahoo! Research and currently funded by Microsoft Research. Microsoft Azure offers many Machine Learning options which VW is one of them. It has many features such as optimization algorithms, reduction functions, importance weighting, and selection of different loss functions. VW is properly runnable in single machines, HPC clusters, and Hadoop.

Strengths

- Efficient, scalable, fast, out-of-core, and open-source online learning supported by competent companies (Microsoft, formerly Yahoo).
- Using a hash via 32-bit MurmurHash3, feature identities are transformed to a weighted index Hashing trick [115] or figure hashing is a fast and resourceful approach to vectorize features that enable rapid online learning.
- VW parses input and exploits multi-core CPUs on Hadoop clusters by its MPI-AllReduce library along with learning, all done in separate threads.
- Enables the use of non-linear features such as n-grams.
- It is one of the offered Machine Learning options by Microsoft Azure.

Weaknesses

- Enough number Machine Learning methods in hand but limited

XGBoost

XGboost is an optimized distributed gradient boosting designed for high efficiency, flexibility, and portability [116]. It is an open-source library implementing the gradient boosting decision tree algorithm. It has become more popular recently since it was chosen by many winning Machine Learning teams in a lot of Machine Learning competitions. XGBoost uses the Gradient Boosting framework for implementing Machine Learning algorithms. It offers a parallel tree boosting (also known as GBDT or GBM), which solves several data science problems expeditiously. The same code runs on main distributed environments (MPI, SGE, Hadoop) and can solve a huge number of problems. The term gradient boosting comes from the idea of improving or boosting a weak model by mixing it with many other weak models hoping the result of generating an overall strong model. XGBoost boosts weak learning models using iterative learning. It offers interfaces for Python, Java, C++, Julia, and R and is compatible with Linux, Windows, and MacOS while supporting GPUs. XGBoost also supports distributed processing frameworks such as Apache Hadoop/Spark/Flink and Dataflow.

Strengths

- Fast execution and model performance
- Parallelization of tree construction while using all of the CPU cores during training
- Distributed computing for training huge models using a cluster of machines
- Out-of-core computation for massive datasets unable to fit into memory
- Cache optimization of data structures and algorithms to exploit hardware in the best way

Weaknesses

- XGBoost is a boosting library designed for tabular data. Thus, it is useless for other tasks such as computer vision or NLP.

Interactive data analytic and visualization tools

To simplify the understanding of difficult concepts and to support decision-makers, analytics results are displayed interactively. Many data visualization packages at different levels of abstraction exist in Python or R, such as bokeh, seaborn, matplotlib, ggplot, and plotly.

In the last year, web-based notebooks/applications have become popular. By integration with data analytic environments, they create and share documents containing equations, visualizations, narrative text, and data-driven live code. The most famous ones are:

- Jupyter notebook (previously iPython notebook) is an open-source application that offers, e.g., statistical modeling, numerical simulations, creation and sharing documents (notebooks), source equations, code, data visualization, and Machine Learning. JupyterLab has recently been launched and aims to improve it.

- Zeppelin is an interactive notebook designed for the processing, analysis, and visualization of large datasets. Native support for Apache Spark distributed computing is supported by Zeppelin. Zeppelin allows prolonging their functionality through different interpreters, namely Python, Spark, SparkSQL, Scala, and shell from Apache Spark analytics platform.

Deep learning frameworks and libraries

Some well-known Machine Learning frameworks and libraries already provide the capabilities to use GPU accelerators for speeding up the learning procedure with supported interfaces. Some of them also allow using optimized libraries like CUDA (cuDNN) and OpenCL to improve performance. The key characteristic of many-core accelerators is that their highly parallel architecture lets them facilitate computations involving matrix-based operations. Software development in the Machine Learning direction community is extremely dynamic and has different layers of abstraction.

TensorFlow

TensorFlow is an open-source software library designed for numerical computation using data flow graphs. TensorFlow was created and is maintained by the Google Brain team in Google's Machine Intelligence research organization for Machine Learning and Deep Learning. It is currently available under the Apache 2.0 open-source license.

TensorFlow is developed for large-scale distributed training and inference. Graph edges represent the multidimensional data arrays (tensors), while the nodes represent mathematical operations communicating the tensors. The distributed TensorFlow architecture consists of distributed master and worker services with kernel implementations.

These services consist of 200 standard operations such as array manipulation, mathematical, control flow, and state management written in C++. TensorFlow was devised to be used both in development/production systems and research. It is runnable on single CPU systems, GPUs, mobile devices, and large-scale distributed systems with hundreds of nodes. Moreover,

TensorFlow Lite is a lightweight alternative for embedded and mobile devices. It allows on-device, low latency Machine Learning inference with a small binary size; however, it covers only a limited set of operators. Hardware acceleration is also supported by TensorFlow Lite using the Android Neural Networks API. TensorFlow programming interfaces consist of APIs for C++ and Python while Go, R, Haskell, and Java are being worked on. TensorFlow is supported in Google and Amazon cloud environments.

Strengths

- Conspicuously the most popular Deep Learning tool
- Open-source, fast-growing
- Supported by a noble industrial company (Google)
- Has a numerical library for dataflow programming, providing the basis for Deep Learning research and development.
- Works with mathematical expressions dealing with multi-dimensional arrays proficiently.
- GPU/CPU computing, mobile computing, powerful in multi-GPU settings, and high scalability of computation across machines and massive datasets.

Weaknesses

- The lower-level API is still hard to use directly to create Deep Learning Models
- Every computational flow must be formed as a static graph, though the TensorFlow Fold package seeks to solve this problem.

Keras

Keras is a Python wrapper library offering links to other Deep Learning tools, namely Theano, CNTK, beta version with MXNet, Deeplearning4j, and TensorFlow. It was developed to enable fast experimentation. It is released under the MIT license. Keras can execute on GPUs, and CPUs were given the implied frameworks perfectly. Keras is developed and maintained by Francois Chollet and has four guiding principles:

1. User-friendliness and minimalism User experience is an important factor in the development of Keras. It follows best practices for lowering a cerebral load by providing consistent and simple APIs.

2. Modularity A Model is assumed as a sequence of graphs or an independent graph, completely configurable modules pluggable together with the least restrictions. Specifically, cost functions, initialization schemes, neural layers, regularization schemes, and activation functions are all self-standing modules to combine and create new models.

3. Easy extensibility New modules are easy to add, and existing modules provide enough examples enabling expressiveness reduction

4. Work with Python Models are described using Python code, which is dense, easy to debug, and simplifying extensibility

Strengths

- Open-source, fast-growing with backend tools from noble industrial companies, namely Google and Microsoft
- Popular API for Deep Learning beside good documentation
- Appropriate way to quickly define Deep Learning Models based up-to-date with backends (e.g., TensorFlow, Theano, CNTK). Keras wraps backend libraries which results in their capabilities being abstracted, and complexity is hidden.

Weaknesses

- There is a trade-off between modularity/simplicity and flexibility, which is not optimal for researching new architectures.
- Multi-GPU is not 100% working yet regarding efficiency and simplicity, as mentioned by some benchmarks that used it with a TensorFlow backend.

Microsoft CNTK

Microsoft Cognitive Toolkit is a commercial distributed Deep Learning framework with large-scale datasets provided by Microsoft Research. CNTK implements productive DNNs trained for image, speech, handwriting, and text data. Its network is like a symbolic graph of vector operations, namely convolution with building blocks or matrix add/multiply. CNTK supports RNN, FFNN, and CNN architectures and implements stochastic gradient descent (SGD) learning with differentiation and parallelization across multiple GPUs and servers. CNTK is compatible with 64-bit Linux and Windows operating systems using C++, C#, BrainScript API, and Python.

Strengths

- Open source, rapidly growing and supported by a noble industrial company (Microsoft)
- Supporting the Open Neural Network Exchange (ONNX) format— co-developed by Microsoft and Facebook—enables easily converting models between CNTK, MXNet, PyTorch, Caffe2, and other Deep Learning tools.
- Faster in performance than Theano and TensorFlow, reporting several benchmarks that used it as Keras backend on multiple machines for LTSM/ RNN.

Weaknesses

- Confined capability on mobile devices

Caffe

Caffe is one of the Deep Learning frameworks designed in respect of speed, modularity, and expression. Yangqing Jia develops it at BAIR (Berkeley Artificial Intelligence Research) along with community contributors. Caffe perceives DNNs layer-by-layer. The layer is the core of a model and the main unit of computation. Data is represented to Caffe through data layers. Proper data sources are efficient databases (LevelDB or LMDB), typical image formats (e.g., JPEG, GIF, PNG, TIFF, PDF), or Hierarchical Data Format (HDF5). Ordinary normalization layers offer different data vector processing and normalization operations. While custom layers—which are less efficient—are also supported in Python, C++ CUDA is necessary for writing new layers.

Strengths

- Appropriate for image processing with CNNs.
- Available pre-trained networks in the Caffe Model Zoo for refinement
- Simple Coding (API/CLI) with Python and MatLab interface

Weaknesses

- It is not being developed as actively as before.
- Many RNN applications that need inconsistent sized inputs are not compatible with static model graph definition
- Defining models in Caffe prototxt files is unhandy for very deep and modular DNN models, namely GoogleLeNet or ResNet, compared to other frameworks.
- Custom layers must be written in C++.

Caffe2

Caffe 2 is a modular, lightweight, and Scalable Deep Learning framework developed by Yangqing Jia and his team at Facebook. Caffe 2 is used at the production level at Facebook while developed in PyTorch. However, it intends to offer a simple way to experiment with Deep Learning and community contributions of new models and algorithms. It differs from Caffe in many enhancement directions, such as adding mobile deployment and new hardware support (CPU and CUDA). Caffe2 is aiming for industrial-strength applications with an excessive concentration on mobile. The basic computation unit in Caffe2 is the operator, which is a more adaptable version of Caffe's layer. More than 400 different operators are at hand in Caffe2, and the community's implementation is anticipated.

Caffe2 provides command-line Python scripts able to translate existing Caffe models into the Caffe2. Regardless, the transformation procedure has to perform a manual verification of accuracy and loss rates. Converting Torch models to Caffe2 models is also possible via Caffe.

Strengths

- Cross-platform, concentrated on mobile devices and edge device inference deployment framework of choice for Facebook
- Supported by Amazon, Intel, NVIDIA, and Qualcomm because of its well-made scalable character in the production
- Enables simply converting models between Caffe2, MXNet, CNTK, PyTorch, and other Deep Learning tools by supporting the Open Neural Network Exchange (ONNX) format.

Weaknesses

- More difficult for Deep Learning beginners compared to PyTorch
- Lack of dynamic graph computation

Torch

Torch is a scientific computing framework widely supporting Machine Learning algorithms based on the Lua programming language. It has been under ongoing development since 2002 [117]. Torch is available under a BSD license for free, and it is supported and used by Facebook, Google, Twitter, DeepMind, and several other organizations. It is implemented in C++ and uses an object-oriented paradigm. Currently, its API is also written in Lua, which is used as a wrapper for optimized for CUDA and C/C++ code. The core of Torch is made up of the Tensor library accessible both with CPU and GPU backends. Several classic operations (such as linear algebra operations) are provided by the Tensor library, which is efficiently implement in C, leveraging SSE instructions on Intel platforms and purposely linking algebra operations to present efficient BLAS/Lapack implementations (such as IntelMKL). Torch supports parallelism on multi-core CPUs by using OpenMP and on GPUs via CUDA. It is generally used for large-scale learning (image, speech, and video applications), supervised learning, unsupervised learning, Neural Networks, image processing, optimization, and graphical models.

Strengths

- Flexible and readable mid-level code beside high-level (Lua)
- Simple code reusability
- Modular and fast
- Extremely suitable for research

Weaknesses

- Still smaller share of projects than Caffe
- LuaJIT is not conventional and causes integration issues
- Despite easy to learn, Lua is not popular
- Ceased development

PyTorch

PyTorch is a Python library for GPU-accelerated Deep Learning. It is a Python interface of the same optimized libraries in C which Torch uses. It has been developed by Facebook's AI research group starting in 2016. PyTorch is written in C, CUDA, and Python. The library binds acceleration libraries like IntelMKL and NVIDIA (cuDNN, NCCL). It uses CPU and GPU Tensor and Neural Network backends (TH, THC, THNN, THCUNN) at the core, written as independent libraries on a C99 API. PyTorch supports tensor computation with mighty GPU acceleration and DNNs built on a tape-based auto grad system. By allowing complicated architectures to be built easily, it has become popular. Generally, when the way a network functions is changed, everything needs to make a new beginning. A technique used by PyTorch called *reverse-mode auto-differentiation* enables changing the way a network functions with the small endeavor (i.e., *dynamic computational graph* or DCG). It is mainly inspired by Chainer and auto grad. The library is available under a BSD license for free and is supported by NVIDIA, Twitter, Facebook, and several other organizations.

Strengths

- Dynamic computational graph (reverse-mode auto-differentiation).
- Supports automatic differentiation for NumPy and SciPy.
- Inventive and flexible Python programming for development.
- Enables simply converting models between Caffe2, MXNet, CNTK, PyTorch, and other Deep Learning tools by supporting the Open Neural Network Exchange (ONNX) format.

MXNet

Apache MXNet is a Deep Learning framework developed with the goal of efficiency and flexibility [118]. Developed by Pedro Domingos and a team of researchers at the University of Washington and a part of the DMLC. It allows the combination of symbolic and imperative programming to enhance efficiency and productivity. MXNet Contains a *dynamic dependency scheduler* at its core which automatically parallelizes both symbolic and imperative operations impulsively. A graph optimization layer over the scheduler makes symbolic execution fast and resourceful. MXNet is portable and lightweight, scaling adequately to multiple GPUs and machines. It also supports an efficient deployment of trained models in substandard devices for inference, namely mobile devices (using Amalgamation), IoT devices (using AWS Greengrass), Serverless (using AWS Lambda), or containers.

MXNet had a wide API language support for Julia, Python, R, and other languages and is licensed under an Apache-2.0 license. Major public cloud providers support MXNet.

Strengths

- Auto parallelism (Dynamic dependency scheduler)
- Outstanding computational scalability with multiple GPUs and CPUs, making it very advantageous for enterprises.
- Supports a flexible programming model and multiple languages, i.e., C++, JavaScript, Go, R, Python, Julia, Matlab, Scala, Perl, and Wolfram.
- Enables simply converting models between Caffe2, MXNet, CNTK, PyTorch, and other Deep Learning tools by supporting the Open Neural Network Exchange (ONNX) format.

Weaknesses

- Occasionally, APIs are not very user-friendly.

Chainer

Chainer is a Python-based, independent open-source framework for Deep Learning models. Its major developers work at Preferred Networks, Inc., a Machine Learning startup with engineers mostly from the University of Tokyo. A wide range of Deep Learning models, including RNN, CNN, reinforcement learning, and variational autoencoders, is provided by Chainer.

Chainer's vision is going further than invariance [119]. Chainer provides automatic differentiation APIs based on Define-by-Run's approach, i.e., *dynamic computational graphs* (DGG), along with high-level object-oriented APIs for building and training Neural Networks. Chainer builds Neural Networks dynamically (computational graph is built impulsively), while other frameworks (such as TensorFlow or Caffe) construct their graph in respect of the *Define-and-Run* scheme (graph is built at the beginning and remains fixed). Chainer supports CUDA/cuDNN using CuPy, to obtain high-performance training and inference. The Intel Math Kernel Library (IntelMKL) for Deep Neural Networks (MKL-DNN), which speeds up Deep Learning frameworks on Intel-based architectures. It also includes libraries for industrial applications such as ChainerCV (for computer vision), ChainerRL (for deep reinforcement learning), and ChainerMN (for scalable multi-node distributed DL).

Strengths

- Dynamic computational graphs based on the *Define-by-Run* principle
- Providing libraries for industrial applications
- Strong companies such as Toyota, FANUC, and NTT invest in it

Weaknesses

- Does not support higher-order gradients.
- DCG needs to be generated each time for fixed networks.

Theano

Theano is an innovative Deep Learning tool supporting GPU computations. It's an open-source project released under the BSD license, and its development began in 2007. LISA group at the University of Montreal actively maintains it (although no longer developed).

Theano is a compiler for mathematical expressions in Python to convert structures into super-efficient code using NumPy and other efficient native libraries like BLAS and native code to run as fast as possible on CPUs or GPUs. It supports an extension for multi-GPU data parallelism and owns a distributed framework to train models.

Strengths

- Open-source and cross-platform
- Strong numerical library providing the basis for DL research and development
- Symbolic API supports looping control, which makes implementing RNNs efficient

Weaknesses

- Lower-level API, difficult for direct use to create Deep Learning models, although wrappers (like Lasagne or Keras) exist.
- Does not support mobile platforms and other programming APIs.
- No longer under active development

Deep learning wrapper libraries

As mentioned earlier, Keras is a wrapper library for Deep Learning libraries aiming to make low-level implementations hidden. Other wrapper libraries include:

- There are lots of wrappers for TensorFlow. External wrapper packages are Keras, TensorLayer, and TFLearn. Wrappers from Google Deepmind are Sonnet and PrettyTensor. Wrappers within native TensorFlow are TF-Slim, tf.Keras, tf.contrib.learn, and tf.layers.

- Gluon is a wrapper for MXNet. Gluon's API specification tries to enhance DL technology's speed, flexibility, and accessibility for all developers, regardless of their DL framework choice. Gluon is a product by Amazon Web Services and Microsoft AI and released under Apache 2.0 license.

- NVIDIA Digits is a web application for training DNNs for image classification, segmentation, and object detection tasks using Deep Learning backends such as Caffe, Torch, and TensorFlow with a wide variety of image formats and sources with Digits plugins. It simplifies typical Deep Learning tasks, namely managing data, designing and training Neural Networks on multi-GPU systems, real-time performance monitorization with advanced visualizations, and selecting the best performing model for deployment based on the results browser. The tool is mostly interactive (GUI) with several pre-trained models, e.g., AlexNet, GoogLeNet, LeNet, and UNet from the Digits Model Store. Digits are released under the BSD 3-clause license.

- Lasagne is a light library to build and train Neural Networks in Theano with six principles:

Simplicity, Transparency, Modularity, Pragmatism, Restraint, and Focus. Blocks and Pylearn2 are the other wrappers for Theano. The development of DL frameworks and libraries is highly dynamic. It thus makes forecasting who will lead this fast-changing ecosystem is difficult, but we can see two main trends happening in the use of DL frameworks:

1. Using TensorFlow for production and Keras for rapid prototyping and. Google aids this trend.

2. Using Caffe2 for production and PyTorch for prototyping. Facebook sponsors this trend. The extensive number of Deep Learning frameworks makes it difficult to develop tools in one framework and use them in others (framework interoperability). The Open Neural Network Exchange (ONNX) attempts to address this problem by introducing an open ecosystem for interchangeable AI models. ONNX is being co-developed by Microsoft, Amazon, and Facebook as an open-source project, and it will support Caffe2, PyTorch, MXNet, and CNTK at first.

References

[1] Goodfellow, I., Bengio, Y., Courville, A. and Bengio, Y. 2016. *Deep Learning* (Vol. 1, No. 2). Cambridge: MIT Press.

[2] Mor-Yosef, S., Samueloff, A., Modan, B., Navot, D. and Schenker, J.G. 1990. Ranking the risk factors for cesarean: logistic regression analysis of a nationwide study. *Obstetrics and Gynecology*, 75(6): 944–947.

[3] Viola, P. and Jones, M.J. 2004. Robust real-time face detection. *International Journal of Computer Vision*, 57(2): 137–154.

[4] Sutton, R.S. and Barto, A.G. 2018. Reinforcement learning: An introduction. MIT Press.

[5] Wiering, M. and Van Otterlo, M. 2012. Reinforcement learning. *Adaptation, Learning, and Optimization*, 12(3).

[6] Tesauro, G. 1994. TD-Gammon, a self-teaching backgammon program, achieves master-level play. *Neural Computation*, 6(2): 215–219.

[7] Russell, S. and Norvig, P. 2002. Artificial intelligence: A modern approach.

[8] Mitchell, T.M. 1997. Machine learning, 870–877.

[9] Hastie, T., Tibshirani, R. and Friedman, J. 2009. The elements of statistical learning: data mining, inference, and prediction. Springer Science & Business Media; 2009 Aug 26.

[10] Bishop, C.M. 2006. *Pattern Recognition and Machine Learning*. Springer.

[11] Friedman, J.H. 1998. Data mining and statistics: What's the connection? *Computing Science and Statistics*, 29(1): 3–9.

[12] Alpaydin, E. 2020. *Introduction to Machine Learning*. MIT press.

[13] Roscher, R., Bohn, B., Duarte, M.F. and J. Garcke. 2020. Explainable Machine Learning for scientific insights and discoveries. In IEEE Access, 8: 42200–42216.

[14] Van Engelen, J.E. and Hoos, H.H. 2020. A survey on semi-supervised learning. *Machine Learning*, 109(2): 373–440.

[15] Kaelbling, L.P., Littman, M.L. and Moore, A.W. 1996. Reinforcement learning: A survey. *Journal of Artificial Intelligence Research*, 4: 237–285.

[16] Kober, J., Bagnell, J.A. and Peters, J. 2013. Reinforcement learning in robotics: A survey. *The International Journal of Robotics Research*, 32(11): 1238–1274.

[17] Arulkumaran, K., Deisenroth, M.P., Brundage, M. and Bharath, A.A. 2017. Deep reinforcement learning: A brief survey. *IEEE Signal Processing Magazine*, 34(6): 26–38.

[18] Samuel, A.L. 1959. Some studies in machine learning using the game of checkers. *IBM Journal of Research and Development*, 3(3): 210–229.

[19] Nilsson, N.J. 1965. Learning machines.

[20] Duda, R.O. and Hart, P.E. 1973. Pattern recognition and scene analysis.

[21] Grossberg, S. 1976. Adaptive pattern classification and universal recoding: II. Feedback, expectation, olfaction, illusions. *Biological Cybernetics*, 23(4): 187–202.

[22] Harnad, S. 2006. The annotation game: On Turing (1950) on computing, machinery, and intelligence. *In: The Turing Test Sourcebook: Philosophical and Methodological Issues in the Quest for the Thinking Computer*. Kluwer.

[23] Hand, D.J. and Adams, N.M. 2014. Data mining. *Wiley StatsRef: Statistics Reference Online*, pp. 1–7.

[24] Han, J., Kamber, M. and Pei, J. 2011. Data mining concepts and techniques third edition. *The Morgan Kaufmann Series in Data Management Systems*, 5(4): 83–124.

[25] Nocedal, J. and Wright, S. 2006. *Numerical Optimization*. Springer Science & Business Media.

[26] Sra, S., Nowozin, S. and Wright, S.J. (eds.). 2012. *Optimization for Machine Learning*. Mit Press.

[27] Van Calster, B. 2019. Statistics versus machine learning: definitions are interesting (but understanding, methodology, and reporting are more important). *Journal of Clinical Epidemiology*, 116: 137.

[28] Breiman, L. 2001. Statistical modeling: The two cultures (with comments and a rejoinder by the author). *Statistical Science*, 16(3): 199–231.

[29] James, G., Witten, D., Hastie, T. and Tibshirani, R. 2013. *An Introduction to Statistical Learning* (Vol. 112, p. 18). New York: Springer.

[30] Mohri, M., Rostamizadeh, A. and Talwalkar, A. 2018. *Foundations of Machine Learning*. MIT Press.

[31] Zhuang, Y., Rui, Y., Huang, T.S. and Mehrotra, S. 1998, October. Adaptive key frame extraction using unsupervised clustering. pp. 866–870. *In: Proceedings 1998 International Conference on Image Processing. icip98 (cat. no. 98cb36269)* (Vol. 1). IEEE.

[32] Woergoetter, F. and Porr, B. 2008. Reinforcement learning. *Scholarpedia*, 3(3): 1448.

[33] Van Otterlo, M. and Wiering, M. 2012. Reinforcement learning and markov decision processes. pp. 3–42. *In: Reinforcement Learning*. Springer, Berlin, Heidelberg.

[34] Bozinovska, S.B.L. 2001. Self-learning agents: A connectionist theory of emotion based on crossbar value judgment. *Cybernetics & Systems*, 32(6): 637–669.

[35] Bozinovski, S. 2014. Modeling mechanisms of cognition-emotion interaction in Artificial Neural Networks, since 1981. *Procedia Computer Science*, 41: 255–263.

[36] Bozinovski, S. 1982. A self-learning system using secondary reinforcement. *Cybernetics and Systems Research*, pp. 397–402.

[37] Hyvärinen, A. 1999. Survey on independent component analysis.

[38] Baldi, P. 2012. June. Autoencoders, unsupervised learning, and deep architectures. pp. 37–49. *In: Proceedings of ICML Workshop on Unsupervised and Transfer Learning*. JMLR Workshop and Conference Proceedings.

[39] Mnih, A. and Salakhutdinov, R.R. 2007. Probabilistic matrix factorization. *Advances in Neural Information Processing Systems*, 20: 1257–1264.

[40] Xu, R. and Wunsch, D. 2005. Survey of clustering algorithms. *IEEE Transactions on Neural Networks*, 16(3): 645–678.

[41] Voulodimos, A., Doulamis, N., Doulamis, A. and Protopapadakis, E. 2018. Deep learning for computer vision: A brief review. *Computational Intelligence and Neuroscience*, 2018.

[42] Deng, L., Hinton, G. and Kingsbury, B. 2013. May. New types of deep neural network learning for speech recognition and related applications: An overview. pp. 8599–8603. *In: 2013 IEEE International Conference on Acoustics, Speech and Signal Processing*. IEEE.

[43] Singh, S.P., Kumar, A., Darbari, H., Singh, L., Rastogi, A. and Jain, S. 2017, July. Machine translation using deep learning: An overview. pp. 162–167. *In: 2017 International Conference on Computer, Communications and Electronics (comptelix)*. IEEE.

[44] Lee, H., Grosse, R., Ranganath, R. and Ng, A.Y. 2009, June. Convolutional deep belief networks for scalable unsupervised learning of hierarchical representations. pp. 609–616. *In: Proceedings of the 26th Annual International Conference on Machine Learning*.

[45] Quinlan, J.R. 1990. Decision trees and decision-making. *IEEE Transactions on Systems, Man, and Cybernetics*, 20(2): 339–346.

[46] Ali, J., Khan, R., Ahmad, N. and Maqsood, I. 2012. Random forests and decision trees. *International Journal of Computer Science Issues (IJCSI)*, 9(5): 272.

[47] Cortes, C. and Vapnik, V. 1995. Support-vector networks. *Machine Learning*, 20(3): 273–297.

[48] Margaritis, D. 2003. *Learning Bayesian Network Model Structure from Data*. Carnegie-Mellon Univ Pittsburgh Pa School of Computer Science.

[49] Sivanandam, S.N. and Deepa, S.N. 2008. Genetic algorithms. pp. 15–37. *In: Introduction to Genetic Algorithms*. Springer, Berlin, Heidelberg.

[50] Yang, Q., Liu, Y., Chen, T. and Tong, Y. 2019. Federated machine learning: Concept and applications. *ACM Transactions on Intelligent Systems and Technology (TIST)*, 10(2): 1–19.

[51] LeCun, Y., Bengio, Y. and Hinton, G. 2015. Deep learning. *Nature*, 521(7553): 436–444.

[52] Yan, L.C., Yoshua, B. and Geoffrey, H. 2015. Deep learning. *Nature*, 521(7553): 436–444.

[53] Voulodimos, A., Doulamis, N., Doulamis, A. and Protopapadakis, E. 2018. Deep learning for computer vision: A brief review. *Computational Intelligence and Neuroscience*, 2018.

[54] Deng, L., Hinton, G. and Kingsbury, B. 2013. May. New types of deep neural network learning for speech recognition and related applications: An overview. pp. 8599–8603. *In: 2013 IEEE International Conference on Acoustics, Speech and Signal Processing*. IEEE.

[55] Nguyen, D.T., Alam, F., Ofli, F. and Imran, M. 2017. Automatic image filtering on social networks using deep learning and perceptual hashing during crises. *arXiv preprint arXiv:1704.02602*.

[56] Min, S., Lee, B. and Yoon, S. 2017. Deep learning in bioinformatics. *Briefings in Bioinformatics*, 18(5): 851–869.

[57] Shen, D., Wu, G. and Suk, H.I. 2017. Deep learning in medical image analysis. *Annual Review of Biomedical Engineering*, 19: 221248.

[58] Justesen, N., Bontrager, P., Togelius, J. and Risi, S. 2019. Deep learning for video game playing. *IEEE Transactions on Games*, 12(1): 1–20.

[59] Nabati, M., Navidan, H., Shahbazian, R., Ghorashi, S.A. and Windridge, D. 2020. Using synthetic data to enhance the accuracy of fingerprint-based localization: a deep learning approach. *IEEE Sensors Letters*, 4(4): 1–4.

[60] Salakhutdinov, R. and Hinton, G. 2009, April. Deep boltzmann machines. pp. 448–455. *In*: *Artificial Intelligence and Statistics*. PMLR.

[61] Schmidhuber, J. 2015. Deep learning in neural networks: An overview. *Neural Networks*, 61: 85–117.

[62] Ng, A. 2011. Sparse autoencoder. *CS294A Lecture Notes*, 72(2011): 1–19.

[63] Lee, H., Grosse, R., Ranganath, R. and Ng, A.Y. 2009, June. Convolutional deep belief networks for scalable unsupervised learning of hierarchical representations. pp. 609–616. *In*: *Proceedings of the 26th Annual International Conference on Machine Learning*.

[64] Hornik, K. 1991. Approximation capabilities of multilayer feedforward networks. *Neural Networks*, 4(2): 251–257.

[65] Robert, C. 2014. Machine learning, a probabilistic perspective. 2014: 62–63.

[66] Lu, Z., Pu, H., Wang, F., Hu, Z. and Wang, L. 2017. The expressive power of neural networks: A view from the width. *arXiv preprint arXiv:1709.02540*.

[67] Bengio, Y. 2009. *Learning Deep Architectures for AI*. Now Publishers Inc.

[68] Szegedy, C., Toshev, A. and Erhan, D. 2013. Deep neural networks for object detection.

[69] Rolnick, D. and Tegmark, M. 2017. The power of deeper networks for expressing natural functions. *arXiv preprint arXiv:1705.05502*.

[70] Bengio, Y., Boulanger-Lewandowski, N. and Pascanu, R. 2013, May. Advances in optimizing recurrent networks. pp. 8624–8628. *In*: *2013 IEEE International Conference on Acoustics, Speech and Signal Processing*. IEEE.

[71] Dahl, G.E., Sainath, T.N. and Hinton, G.E. 2013, May. Improving deep neural networks for LVCSR using rectified linear units and dropout. pp. 8609–8613. *In*: *2013 IEEE International Conference on Acoustics, Speech and Signal Processing*. IEEE.

[72] Hinton, G.E. 2012. A practical guide to training restricted Boltzmann machines. pp. 599–619. *In*: *Neural Networks: Tricks of the Trade*. Springer, Berlin, Heidelberg.

[73] Ding, S., Zhang, N., Xu, X., Guo, L. and Zhang, J. 2015. Deep extreme learning machine and its application in EEG classification. *Mathematical Problems in Engineering*, 2015.

[74] Huang, G., Huang, G.B., Song, S. and You, K. 2015. Trends in extreme learning machines: A review. *Neural Networks*, 61: 32–48.

[75] Albawi, S., Mohammed, T.A. and Al-Zawi, S. 2017, August. Understanding of a convolutional neural network. pp. 1–6. *In*: *2017 International Conference on Engineering and Technology (ICET)*. IEEE.

[76] Hochreiter, S. and Schmidhuber, J. 1997. Long short-term memory. *Neural Computation*, 9(8): 1735–1780.

[77] Mandic, D. and Chambers, J. 2001. *Recurrent Neural Networks for Prediction: Learning Algorithms, Architectures and Stability*. Wiley.

[78] Goodfellow, I. J., Pouget-Abadie, J., Mirza, M., Xu, B., Warde-Farley, D., Ozair, S., Courville, A.C. and Bengio, Y. 2014. Generative adversarial nets. pp. 2672–2680. *In*: *Proceedings of NIPS*.

[79] Park, J. and Sandberg, I.W. 1993. Approximation and radial-basis-function networks. *Neural Computation*, 5(2): 305–316.

[80] Baum, E.B. 1988. On the capabilities of multilayer perceptrons. *Journal of Complexity*, 4(3): 193–215.

[81] Hinton, G.E. 2009. Deep belief networks. *Scholarpedia*, 4(5): 5947.

[82] Hinton, G.E. 2012. A practical guide to training restricted Boltzmann machines. pp. 599–619. *In: Neural Networks: Tricks of the Trade*. Springer, Berlin, Heidelberg.

[83] Baldi, P. 2012, June. Autoencoders, unsupervised learning, and deep architectures. pp. 37–49. *In: Proceedings of ICML Workshop on Unsupervised and Transfer Learning*. JMLR Workshop and Conference Proceedings.

[84] Yi, Z., Zhang, H., Tan, P. and Gong, M. 2017. Dualgan: Unsupervised dual learning for image-to-image translation. *In: Proceedings of the IEEE International Conference on Computer Vision.*

[85] Zhu, L., Chen, Y., Ghamisi, P. and Benediktsson, J.A. 2018. Generative adversarial networks for hyperspectral image classification. *IEEE Transactions on Geoscience and Remote Sensing*, 56(9): 5046–5063.

[86] Zhang, Y., Gan, Z. and Carin, L. 2016. Generating text via adversarial training. *In: NIPS Workshop on Adversarial Training* (Vol. 21).

[87] Pascual, S., Bonafonte, A. and Serra, J. 2017. SEGAN: Speech enhancement generative adversarial network. *arXiv preprint arXiv:1703.09452.*

[88] Hu, W.W. and Tan, Y. 2017. Generating adversarial malware examples for black-box attacks based on GAN. *arXiv: 1702.05983.*

[89] Vondrick, C., Pirsiavash, C.H. and Torralba, A. 2016. Generating videos with scene dynamics. *In: Proceedings of the Conference on Neural Information Processing Systems.*

[90] Liang, X., Lee, L., Dai, W. and Xing, E.P. 2017. Dual motion GAN for future-flow embedded video prediction. *In: Proceedings of the IEEE International Conference on Computer Vision.*

[91] Mirza, M. and Osindero, S. 2014. Conditional generative adversarial nets. *In: J. CoRR.*

[92] Odena, A., Olah, C. and Shlens, J. 2017. Conditional image synthesis with auxiliary classifier GANs.

[93] Arjovsky, M. and Bottou, L. 2017. Towards principled methods for training generative adversarial networks. *arXiv preprint arXiv:1701.04862, 2017.*

[94] Arjovsky, M., Chintala, S. and Bottou, L. 2017. "Wasserstein gan," 2017.

[95] Gulrajani, I., Ahmed, F., Arjovsky, M., Dumoulin, V. and A.C. Courville. 2017. Improved training of wasserstein gans. *In: Advances in Neural Information Processing Systems.*

[96] Chen, X., Duan, Y., Houthooft, R., Schulman, J., Sutskever, I. and Abbeel, P. 2016. Infogan: Interpretable representation learning by information maximizing generative adversarial nets. *In: Advances in Neural Information Processing Systems*, 2016.

[97] Mao, X., Li, Q., Xie, H., Lau, R.Y.K., Wang, Z. and P. Smolley, Stephen. 2017. Least squares generative adversarial networks. *In: Proceedings of the IEEE International Conference on Computer Vision*, 2017.

[98] Donahue, J., Krahenbuhl, P. and Darrell, T. 2016. Adversarial feature learning. *In: arXiv preprint arXiv:1605.09782.*

[99] Unsupervised representation learning with deep convolutional generative adversarial networks. *In: Proceedings of the International Conference on Learning Representations Workshop Track*, 2016.

[100] Jouppi, N.P., Young, C., Patil, N., Patterson, D., Agrawal, G., Bajwa, R., Bates, S., Bhatia, S., Boden, N., Borchers, A. and Boyle, R. 2017, June. In-data center

performance analysis of a tensor processing unit. pp. 1–12. *In: Proceedings of the 44th Annual International Symposium on Computer Architecture.*

[101] Cano, A. 2018. A survey on graphic processing unit computing for large-scale data mining. *Wiley Interdisciplinary Reviews: Data Mining and Knowledge Discovery*, 8(1): e1232.

[102] Nguyen, G., Nguyen, B.M., Tran, D. and Hluchy, L. 2018. A heuristics approach to mine behavioural data logs in mobile malware detection system. *Data & Knowledge Engineering*, 115: 129–151.

[103] Liu, J., Li, J., Li, W. and Wu, J. 2016. Rethinking big data: A review on the data quality and usage issues. *ISPRS Journal of Photogrammetry and Remote Sensing*, 115: 134–142.

[104] Lacey, G., Taylor, G.W. and Areibi, S. 2016. Deep learning on fpgas: Past, present, and future. *arXiv preprint arXiv:1602.04283.*

[105] Courbariaux, M., Hubara, I., Soudry, D., El-Yaniv, R. and Bengio, Y. 2016. Binarized neural networks: Training deep neural networks with weights and activations constrained to + 1 or −1. *arXiv preprint arXiv:1602.02830.*

[106] Iandola, F.N., Han, S., Moskewicz, M.W., Ashraf, K., Dally, W.J. and Keutzer, K. 2016. SqueezeNet: AlexNet-level accuracy with 50x fewer parameters and < 0.5 MB model size. *arXiv preprint arXiv:1602.07360.*

[107] Hermans, J. 2017. *On Scalable Deep Learning and Parallelizing Gradient Descent* (Doctoral dissertation, Maastricht U.).

[108] Krizhevsky, A. 2014. One weird trick for parallelizing convolutional neural networks. *arXiv preprint arXiv:1404.5997.*

[109] Jovic, A., Brkic, K. and Bogunovic, N. 2014, May. An overview of free software tools for general data mining. pp. 1112–1117. *In: 2014 37th International Convention on Information and Communication Technology, Electronics and Microelectronics (MIPRO).* IEEE.

[110] Sonnenburg, S., Rätsch, G., Henschel, S., Widmer, C., Behr, J., Zien, A., Bona, F.D., Binder, A., Gehl, C. and Franc, V. 2010. The SHOGUN machine learning toolbox. *The Journal of Machine Learning Research*, 11: 1799–1802.

[111] Mierswa, I., Klinkenberg, R., Fischer, S. and Ritthoff, O. 2003, August. A flexible platform for knowledge discovery experiments: Yale–yet another learning environment. *In: LLWA 03-Tagungsband der GI-Workshop-Woche Lernen-Lehren-Wissen-Adaptivität.*

[112] Abdiansah, A. and Wardoyo, R. 2015. Time complexity analysis of support vector machines (SVM) in LibSVM. *International Journal Computer and Application*, 128(3): 28–34.

[113] Fan, R.E., Chang, K.W., Hsieh, C.J., Wang, X.R. and Lin, C.J. 2008. LIBLINEAR: A library for large linear classification. *The Journal of Machine Learning Research*, 9: 1871–1874.

[114] Crammer, K. and Singer, Y. 2001. On the algorithmic implementation of multiclass kernel-based vector machines. *Journal of Machine Learning Research*, 2(Dec): 265–292.

[115] Weinberger, K., Dasgupta, A., Langford, J., Smola, A. and Attenberg, J. 2009, June. Feature hashing for large scale multitask learning. pp. 1113–1120. *In: Proceedings of the 26th Annual International Conference on Machine Learning.*

[116] Chen, T. and Guestrin, C. 2016, August. Xgboost: A scalable tree boosting system. pp. 785–794. *In*: *Proceedings of the 22nd acm sigkdd International Conference on Knowledge Discovery and Data Mining.*

[117] Collobert, R., Bengio, S. and Mariéthoz, J. 2002. *Torch: A Modular Machine Learning Software Library* (No. REP_WORK). Idiap.

[118] Chen, T., Li, M., Li, Y., Lin, M., Wang, N., Wang, M., Xiao, T., Xu, B., Zhang, C. and Zhang, Z. 2015. Mxnet: A flexible and efficient machine learning library for heterogeneous distributed systems. *arXiv preprint arXiv:1512.01274.*

[119] Tokui, S., Oono, K., Hido, S. and Clayton, J. 2015. December. Chainer: a next-generation open source framework for deep learning. pp. 1–6. *In*: *Proceedings of Workshop on Machine Learning Systems (LearningSys) in the Twenty-Ninth Annual Conference on Neural Information Processing Systems (NIPS)* (Vol. 5).

[120] Jagtap, A.D., Kawaguchi, K. and Karniadakis, G.E. 2020. Adaptive activation functions accelerate convergence in deep and physics-informed neural networks. *Journal of Computational Physics*, 404: 109136.

Index

||||||||||||||||||||||||||||||||||||||||||||||||||||||||||||||||||||||||||||||||||||||||||||||||||||||||||||||||||||||||||||||||||||||||||||||||||||||||||||||||||||||||||||||||||||||